FROM LANCE TO LANDIS

FROM LANCE TO LANDIS

Inside the American Doping Controversy at the Tour de France

DAVID WALSH

Ballantine Books New York

Published in the United States by Ballantine Books, an imprint of The Random House Publishing Group, a division of Random House, Inc., New York.

ISBN 978-0-345-49962-2

Printed in the United States of America on acid-free paper

www.ballantinebooks.com

246897531

First Edition

Book design by Carol Malcolm Russo

For John

CONTENTS

Glossary ix

Dramatis Personae xi

Prologue xiii

Chapter One: **The Kid from the Cornfields** 3

Chapter Two: **The Needle and the Damage Done** 13

Chapter Three: **New Kid, Old World** 28

Chapter Four: **The Terrible Elixir** 39

Chapter Five: **If You Can't Beat Them . . .** 63

Chapter Six: **The Hospital Room—Part One** 73

Chapter Seven: **Postal Goes European** 83

Chapter Eight: **The Leader Returns** 96

Chapter Nine: **The Program** 111

Chapter Ten: **Crossing the Line** 121

Chapter Eleven: ***Plus Ça Change, Plus C'est la Même Chose* (The More Things Change, the More They Stay the Same)** 138

Chapter Twelve: **Frankie's Breaking Point** 154

Chapter Thirteen: **A Strange Kind of Glory** 165

Chapter Fourteen: **LeMond Feels the Heat** 178

Chapter Fifteen: **The Strange World of Ty** 192

Chapter Sixteen: **The Empire Strikes Back** 210

Chapter Seventeen: **The Sting in the Tale** 226

Chapter Eighteen: **The Hospital Room—Part Two** 248

Chapter Nineteen: **One in a Billion?** 272

Chapter Twenty: **"I'll Say No"** 293

Epilogue: **The Man Is More Than the Cyclist** 319

Author's Note 331

GLOSSARY

blood doping: the manipulation of blood within a person's body by means of drugs or transfusions.

carbon isotope test: a sophisticated test that can identify synthetic testosterone in urine.

CAS: Court of Arbitration for Sport; hears athletes' appeals against doping convictions.

centrifuge: an apparatus consisting of a compartment that is spun about a central axis; can be used to measure hematocrit.

CONI: the Italian Olympic Committee.

corticosteroids/corticoids: steroid hormone produced by the adrenal cortex or synthesized; without a therapeutic exemption, presence of synthesized corticosteroids/corticoids in an athlete's body constitutes doping.

directeur sportif: sports director of a professional cycling team.

équipier: a professional cyclist who rides for the team and its leader.

hematocrit: the amount of red blood cells in the blood expressed as a percentage of total blood volume.

human growth hormone: often written as HGH; widely abused as a performance-enhancing drug.

LNDD: Laboratoire National de Dépistage du Dopage; French national anti-doping laboratory.

passive doping: the suffering and demoralization experienced by clean athletes trying to compete against doped rivals.

Pot Belge: the slang name given to a cocktail of recreational drugs abused by some professional cyclists.

r-EPO: recombinant erythropoietin; a drug that stimulates the production of red blood cells.

soigneur: literally means "carer"; the name given to a masseur or masseuse on a cycling team.

testosterone: a hormone abused by athletes in its synthetic form.

UCI: Union Cycliste Internationale; the world body that governs the sport of cycling.

USAC: USA Cycling; formerly USCF, United States Cycling Federation.

USADA: United States Anti-Doping Agency.

WADA: World Anti-Doping Agency.

DRAMATIS PERSONAE

Lance Armstrong: retired seven-time winner of the Tour de France

Dr. Greg Strock: former elite-level U.S. amateur cyclist

René Wenzel: Greg Strock's one-time coach

George Hincapie: professional cyclist, longtime teammate of Armstrong, currently riding with the Discovery team

Kevin Livingston: former professional cyclist and teammate of Armstrong

Greg LeMond: former three-time winner of, and first American to win, the Tour de France

Frankie Andreu: former racer, and former friend and teammate of Armstrong

Betsy Andreu: wife of Frankie Andreu

Andy Hampsten: former racer and teammate of LeMond and, later, Armstrong

Professor Francesco Conconi: controversial Italian doctor, accused of doping athletes

Dr. Michele Ferrari: former protégé of Professor Conconi, onetime trainer of Armstrong, also accused of doping athletes

Hein Verbruggen: former president of the world cycling authority, Union Cycliste Internationale

Stephen Swart: former racer and onetime teammate of Armstrong on Motorola team

Emma O'Reilly: former head soigneur for the the U.S. Postal Service team, and former masseuse of Armstrong

Tyler Hamilton: onetime teammate of Armstrong on the U.S. Postal Service team, subsequently banned for two years following doping offense

Dr. Prentice Steffen: former doctor for the U.S. Postal Service team

Jonathan Vaughters: former racer and former teammate of Armstrong on the U.S. Postal Service team

Stephanie McIlvain: Oakley sportswear employee responsible for direct relationship with Armstrong

Dr. Michael Ashenden: Australian anti-doping expert

Floyd Landis: winner of 2006 Tour de France, subsequently accused of cheating

PROLOGUE

Monday, July 19, 1999. It is a rest day on the Tour de France and the race's three-thousand-strong entourage has set up camp in the Pyrenean town of Saint-Gaudens. After two weeks on the road, it is an opportunity for the Tour's traveling community to draw breath before the final push north to Paris. For Benoît Hopquin, a journalist with the French daily newspaper *Le Monde,* it is another day at the office. He attends the press conference of the champion-elect Lance Armstrong and is at the pressroom later in the afternoon when he takes a phone call from a source in Paris. They disagree over something Hopquin wrote a few days before, but things soon cool down and they talk about the Tour. During the first week of the race there had been a story about riders testing positive for corticoids and a rumor that Armstrong had been one. Hopquin tells his source about the press conference and Armstrong's insistence that he has never used corticoids and didn't have a medical exemption for any banned product. "*Ce n'est pas vrai*" (It's not true), says the source, teasingly, because he is in a position to know. "What are you saying?" asks Hopquin. The source refuses to elaborate, preferring to leave the journalist with an impression. Hopquin's impression is that Armstrong *has* tested positive for corticoids.

Hopquin spoke with his *Le Monde* colleagues, Yves Bordenave and Philippe Le Coeur, and they began calling contacts who might know the truth. They rang a source at the French Ministry of Youth and Sport, they called the national anti-doping laboratory at Châtenay-Malabry, they left two messages for the head of the medical commission of Union Cycliste Internationale (UCI), Leon Schattenberg, but they couldn't confirm the story. They tracked down UCI president Hein Verbruggen, who said he had not been advised about a positive test for Armstrong. The following morning, the journalists went back to Hopquin's original source. He agreed to meet a representative of the newspaper and to show him the medical report forms of the riders who had tested positive for corticoids. A senior editor from *Le Monde* went to the rendezvous and saw a medical report for Armstrong that showed he had tested positive for corticoids following a drug test on the first weekend of the race. The journalist looked down the form to the part where it said *Médicaments Pris* (medications taken), and the word *néant* (none) was written. *Le Monde* believed it was on to an important story.

From the newspaper's point of view, Armstrong's insistence that he did not have a therapeutic exemption was critical. Corticoids are banned but may be legally taken if supported by a doctor's prescription. At the previous day's press conference and in an answer to a journalist from *L'Equipe* eleven days before, Armstrong categorically said he did not have any medical exemptions. Journalists from *Le Monde* contacted U.S. Postal Service team spokesman Dan Osipow and were told the team would wait for an official declaration from UCI before making any comment.

Twelve days before, the U.S. Postal Service team had been told about the presence of cortisone in Armstrong's drug test. Selected at random, Postal rider Kevin Livingston was accompanied to the medical caravan by the team's directeur sportif, Johan Bruyneel. There, the two men heard about the positive test, and then for twelve days, there was nothing, until *Le Monde*'s journalists started asking questions. On the day after the rest day, the Tour riders left Saint-Gaudens on the race to the Pyrenean ski station at Piau-Engaly. It was a tough leg, won by the Spaniard Fernando Escartin, and it was also a good day for race leader Armstrong. He crossed the line fourth and tightened his grip on the race. That day, *Le Monde* published its story alleging he had tested positive for the corticoid triamcinolone acetonide. If the story were substantiated, Armstrong's dream of winning the Tour de France would be over.

That evening Armstrong traveled by helicopter from the top of the mountain to the team hotel. By the time he got around to his evening massage, it was late and two high-ranking U.S. Postal Service team officials were in the room. They spoke with the rider about *Le Monde*'s story and discussed how they would counter it.

At stake was the greatest comeback story in the history of sport.

FROM LANCE TO LANDIS

Chapter 1

THE KID FROM THE CORNFIELDS

It was one of the tougher moments in Greg Strock's unfulfilled career in cycling: the moment when he had to accept it was over. The dream of becoming a professional cyclist had ended much too soon. He was just twenty-one. At the same time, the 1993 Tour de France was winding its way south toward the Alps, and he was in Madison, Wisconsin, riding among fellow Americans in a race that slipped under the sport's radar. Though he now felt a long way from the elite peloton, there had been a time when Strock imagined himself among them. At the age of seventeen, he had been offered a place in the amateur squad of Spain's Banesto team. He thought he would go there, impress the locals, and earn a place on a top professional team. But that was then, before illness sucked away his energy, drained away his ambition.

He waited almost two years for his body to recover. Ever so

slowly, it did. It improved enough, anyway, for him to feel normal and to try to resurrect his career. And though his second coming had its moments, in the end he couldn't get back to where he'd been. One day he felt strong, rode well, believed it was possible. The next morning his body spoke to him, not so much of aches and pains but of overwhelming tiredness. To be successful, a cyclist needs to recover fast. Now Strock knew he wasn't going to be a successful racer, and on that July afternoon in Madison, he let it go. "I can't do this anymore," he said to his coach, René Wenzel, on a street not far from the finish line. "I'm going to go to medical school, because I can't do this anymore."

This wasn't what Wenzel wanted to hear. Part of the reason he had taken the assistant team director's job with Saturn was so he could work again with Strock. He'd had him as a junior in the U.S. squad three years before and liked him. Though Strock was now telling him it was over, he didn't accept that. What's wrong with persevering? "This will come right," he told Strock. "You can still get back. You can be very good again." What began as a heart-to-heart conversation ended as a full-blown argument. Wenzel yelled as Strock stuck by his decision. In the midst of the coach's accusations and the rider's stubbornness, Wenzel saw tears well in Strock's eyes, and that made him emotional but he didn't want to cry, not in front of his rider, and so he kept shouting. Eventually, emotions calmed, and with that came the certainty that it *was* over. Wenzel had to accept that one of his favorite riders was leaving the sport. That evening they shared a beer and tried to end things the right way.

After that they went their separate ways and over the following years they would drift apart. Accepted into medical school at Indiana University, Strock wasn't left with a lot of free time. Wenzel continued with the Saturn cycling team, got laid off, then got rehired; to better make ends meet, he started his own business, Wenzel Coaching, from his home in McKenzie Bridge, Oregon. Coaching bike riders was something he was good at. But, seven years on from that angst-ridden scene in Wisconsin, Strock reentered his life. Wenzel was in Bermuda with a women's team due to race the Bermuda Grand

Prix when his wife, Kendra, called. "Your buddy Strock is going to sue you," she said.

You could have knocked him over with a feather.

On an afternoon in January 2001, in a Starbucks coffee shop in Indianapolis, a young man with a latte sits at a table. Over the following two hours, he will turn back the pages of his life and recall a remarkable story.

He is now Dr. Greg Strock, having just graduated from Indiana University medical school, and will soon begin his residency. He was born in Anderson, a town northeast of Indianapolis, and his story began as simply as a fairy tale. From his cousin Dan Taylor, he got the longing to ride a racing bike. Taylor would go for thirty miles at a time, and when you're his impressionable twelve-year-old cousin, that seems a long, long way. Perhaps it was the promise of a world beyond his little Midwest town, the allure of cobbled roads in Belgium, or maybe it was just the thrill of speed, the cut and thrust of competition. Whatever it was, the boy dreamed of being a bike racer. To get his first bike, he mowed neighbors' lawns and saved his earnings until they were enough to trade for a bicycle. When his parents wondered if it was right to spend so much time on a bike, the boy worked harder at his schoolwork and earned the right to his bike time.

As his grades improved, so, too, did his parents' attitude toward cycling. Soon there existed an unspoken pact: as long as he took care of his academic work, they made sure he had a lift to the next race, money for the next wheel, and a smiling face to greet him when he returned from a training ride. Adolescent life is simpler when it is controlled by one passion, and back then Greg Strock existed to ride his bike. Everything was arranged so that when the time came, as it did most days of the week, the boy could escape on his bike onto long, straight roads hedged by fields of corn.

He sent away for videos of cycling's biggest races and watched them over and over again, especially the one-day classic race from Paris to Roubaix. True cycling men know this is the hardest of the one-day races, and Strock dreamed of winning it. And all the time, he nurtured his dream. Through the flat cornfield countryside around Anderson, the wind was his constant enemy and, to fight it, he imagined he was chasing breakaways on the cobbled tracks to Roubaix. When the Indiana weather worsened in winter and it was too bad to be outdoors, he rode a stationary bike indoors and took pleasure in the pain. By the time he was fifteen, he was Indiana's best young cyclist and one of the best in the country. That was confirmed a year or so later when he went to the junior national championships at Allentown in Pennsylvania and beat George Hincapie to win the individual time trial. That performance won him a place on the U.S. team for the 1989 junior world championships at Krylatskoye in Moscow, a serious achievement for a boy who would still be a junior in 1990.

At this time cycling in America was run by the United States Cycling Federation (USCF), an organization that would merge with an umbrella organization, USA Cycling (USAC), in 1995. Strock's selection to the team for Moscow meant that he was invited to training camps at Colorado Springs, and he impressed with his performance in physiological tests. "That's better than Roy Knickman in his junior year," they marveled, because Knickman had been very good at that age. They made him do an ergometer test, riding against resistance, and he beat the record for sustained power set by Greg LeMond eleven years before. He ate well, he slept well, he didn't flinch on training, and through it all he didn't feel like he was making a sacrifice. "I was just this kid from the cornfields who found a sport he loved and a sport he was good at."

His experience at the 1989 junior world championships in Moscow sharpened his ambition. He raced in the four-man time-trial team, which didn't suit him, and though they did not get a medal, the experience gave him a taste for serious competition. He returned from the world championships with the belief that he could do a lot better, and after speaking with USCF coaches, it was agreed that for the following season he would compete in the stage races he preferred. Another part of the plan for 1990 was to train even harder, and as he had graduated from high school that January, his parents didn't object. Two and a half years away, the Barcelona Olympics was a target; and after that, he wanted a place on a professional team. So determined was he to make it happen that he agreed to an international exchange with a talented young Spanish cyclist, Igor Gonzalez de Galdeano, which meant he would spend ten weeks racing in Spain during the spring of '90 and Gonzalez de Galdeano would come to the United States for the same amount of time later in the year.

He won three of the six races he rode in Spain, beating many of the best young riders in that country and delivering the performances that got him the offer from Banesto's amateur squad.

So far, so good.

As Strock was packing his suitcases for Spain, the USCF appointed Chris Carmichael and René Wenzel to coach and manage its best amateur racers. Both were ex-riders, and both were young and ambitious. Wenzel thought he was hired to look after the senior amateur squad and that Carmichael would take the juniors, but when he showed up at the USCF offices in Colorado Springs on February 2, 1990, he realized Carmichael was getting the seniors. Carmichael had started the day before, and though Wenzel suspected the change came from a desire to have an American in charge of the more important senior squad, he didn't mind. He would prove himself with the juniors.

He was born René Wenzel Olesen in Copenhagen on April 20, 1960. The middle name was his father's first name and he took the name Olesen from his mother, as his parents were not married. Cycling was in his blood. His father had been a professional cyclist, even if Wenzel Jorgensen's career had been diminished by the onset of the Second World War. Jorgensen retired in 1959, but when young René was ready to ride, his dad was still competing in masters' races and father and son trained together: six- or twelve-mile rides, enough to get the kid into the sport, more than enough to ease the old man out. People saw them together and sensed the kid's longing. They called him Wenzel after his father and the boy didn't fight it. It said

René Wenzel Olesen on his passport but he was known simply as René Wenzel. Eventually, he would officially drop Olesen and his passport would catch up with the reality.

Like the young man he would coach ten years later, when Wenzel was in high school, he dreamed of becoming a professional cyclist. School was okay but not what he wanted, and five months into his final year, he quit and headed for Belgium. Deinze is a Flanders town close to the spiritual heartland of European cycling. He arrived in early 1979, a young Danish kid with a little money and a lot of hunger. At first he competed as an individual, but he rode well enough to be taken on by the local club, KVC (Kronica Velo Club) Deinze. It wasn't the happiest time in his life because he was only nineteen and unused to being away from home; he found all the free time hard. Good results would have helped, but he struggled on the cobbled roads used in so many of the races in Flanders.

Wenzel was not stupid and knew the people who befriended him at his new Belgian club expected him to get results for KVC. If the results didn't come, they would move on to some other wannabe. He also sensed that no one much cared how good results were achieved. If some racer found a pill that worked or accepted that kind of help from a friend or trainer, that was a private matter.

One KVC club member helped the foreign riders in Deinze. Took them from the airport to their lodgings, brought them to races, showed them where to shop, and tried to be their friend. Before one race at Mariakerke on the coast, this man gave the teenager two tablets. "These are good vitamins," he said. Wenzel accepted. "He was my guy," he says now, "the guy that I was with and I didn't think it was doping." Wenzel took the tablets and twenty minutes later felt his heart begin to beat faster, so strong it pounded against his chest wall. Though the Mariakerke race was insanely hard, he was able to keep going. So many times he wanted to quit but something pushed him on and he made it to the finish.

He didn't sleep that night, or the following night. His heart raced on, faster than normal, *thump, thump, thump.* Sometime later, when he understood more about drugs, he realized he had been given amphetamines. The thought of turning down the help never entered his mind. Pills and tablets were part of the sport, and at the time, he was happy to accept they were "just vitamins." After all, it's easier to say yes when you're unsure what you're getting. About the worth of the forbidden fruit, he was of two minds: it had helped, but seventeenth place wasn't much compensation for two sleepless nights. The dream of being a pro survived that season in Belgium, and early in 1980 he left Copenhagen for Paris—another country, another chance. This time the arrangements were better. He rode for a club in the southwest corner of suburban Paris that provided accommodation, a bike, and a small allowance. He and two other Danish riders shared a decent apartment, and the season in France was more enjoyable than his year in Belgium.

Wenzel's experience of cycling's doping culture while in France came when he and his teammates were blood-tested. "We were taken to a doctor," he says, "a man we knew as Dr. Bernard. His full name was Bernard Sainz. We felt like pros when we went to him because if we were being put under medical supervision, that meant someone was taking us seriously. We also knew that Dr. Bernard worked with Bernard Hinault at this time, and that was part of the sell for us. We felt honored that Hinault's doctor would agree to work with us."

In 1980, not much was known about Bernard Sainz. He rode in the '50s and '60s on the track, and after retiring, he worked as a trainer and homeopath with the French GAN-Mercier team. They called him "doctor" even though he had no medical training, and it was also thought that he was a veterinary surgeon but, again, that wasn't true. His relationship with Bernard Hinault was important because in the late 1970s and early '80s, Hinault was the world's best bike rider. In 1978, he won the first of five victories in the Tour de France at the impressively young age of twenty-two. As well as working with Hinault, Sainz had also helped the 1976 winner of the Tour, the Belgian Lucien Van Impe. To their minds, Wenzel and his teammates could not have been taken to a "doctor" with more impressive credentials.

Over the next two decades, official attitudes toward doping and toward those who helped riders dope would change in continental Europe, especially in France. And the police would eventually catch up with Bernard Sainz. In 1986 he was investigated on suspicion of dealing in amphetamines at a Paris track meet, and thirteen years later he was placed under investigation by French authorities on suspicion of breaking anti-doping laws and illegally practicing as a doctor. During this time, he was prohibited from meeting with athletes, attending any cycle races, or having any involvement in the training and preparation of cyclists. He was also ordered not to leave France, and after being stopped in Belgium on a speeding offense (he was visiting the cyclist Frank Vandenbroucke), Sainz was imprisoned upon returning to his native country. In 2005, he was named as a key figure in a major investigation into the doping of racehorses, and though that inquiry is ongoing, Sainz has not been convicted on a doping offense. Philippe Gaumont, a former professional rider who has admitted doping, says that Sainz gave him legal homeopathic medicines to help rid his body of doping products that were supplied by others.

To the French press, he became known as *Dr. Mabuse* after the dark character in the 1922 Fritz Lang movie of the same name. It was an alias that did not cause Sainz excessive concern: he wrote the story of his life in 2002 and called it "The Astonishing Revelations of Doctor Mabuse."

Young and undoubtedly a little naïve, René Wenzel was pleased to have Dr. Bernard examine his blood. The follow-up to the test came a week or so later. "We had finished a training ride one afternoon," Wenzel says, "and those of us who lived in the area were told to stop by at the directeur sportif's house. There was a box containing envelopes with each of our names on them. Inside each envelope was a bunch of tablets, one or two ampoules of liquid, and enough syringes to cover the doses outlined in the note describing what we were to do with the different products. There was also a substance that we were to take, one drop each day placed under the tongue."

When a young rider is given syringes and the hermetically sealed glass container holding what seems a precious—perhaps magical—liquid, and when this substance comes from someone who is described as a doctor and serves as the champion's *preparatore* (preparer), it is natural that he willingly submits. By agreeing, the young rider enters another world and perhaps for the first time sees a picture of sport that is cynical, brutal, and tawdry. Once he accepts the drugs, his view of sport is changed. The virtues that attracted him as a child are overtaken by the reality of what it takes to win. Passion is replaced by a more businesslike attitude, and though success may follow, there can be no realization of the original dream.

"We, the three Danish guys, had no idea," says Wenzel. "We hadn't injected ourselves before and so we ended up injecting each other. The French guys knew exactly what to do. The supply lasted about six weeks. There were tablets we had to take one or two hours before the race, different ones for different races. If I remember correctly the tablets were for the smaller races because there was the possibility of drug controls at the bigger races. In his note telling us how to use the ampoules and tablets, Dr. Bernard didn't say what the products were and there were no labels identifying them. At the time I didn't think of it in terms of doping, but later on, yes."

After returning from France to Denmark in the winter of 1980, Wenzel felt listless and weak. He tried to continue racing but the effort drained his strength and made him want to lie down. For much of the following six months, he was forced to stay in bed. His recovery was slow and he remained sick for a year and a half. Mononucleosis, a debilitating condition caused by an excess of leukocytes (white cells) in the blood, was diagnosed and Wenzel was not able to return to action until the spring of 1982. Even then his body couldn't take a heavy training load and he had to back off at the first hint of fatigue. Eventually he would get back to a high level of racing, but he never returned to where he had been before.

Twenty-six years have passed since Wenzel was an amateur racer in France, and with the passing of time has come clarity. "I would be really surprised if what Bernard Sainz gave us was all legal. If it was, he could have just given us a prescription and sent us to the nearest pharmacy. If it was legal, I think the labels would still have been there. I have no doubt that the shit I was given was probably what made me sick."

Wenzel left Denmark for America in the spring of 1983. His plan was to train in California and then race on the U.S. circuit through that summer. Midway through the season it was obvious his body wasn't ready for the demands of bicycle racing, and for the following two years he raced a little and seesawed between his native country and the United States. By 1986, his health had improved enough to allow a full return to racing and he competed on the U.S. circuit in '86 and '87 before immigration officials caught up with him and suggested his U.S. vacation had gone on for a little too long. But it had been good while it lasted, and soon after he returned to Denmark, Wenzel was offered the job of cycling coach to his old club in Copenhagen. He worked as a postman to supplement his income and kept in touch with American friends in the hope that something might turn up in the United States. In late 1989, it did. Jiri Mainus, coaching director for the United States Cycling Federation, called and asked if he would be interested in a coaching role with his organization. It was precisely what Wenzel wanted, and on February 2, 1990, he walked into the office he would share with fellow coach Chris Carmichael at the USCF's Colorado Springs headquarters in Colorado.

Wenzel's first important assignment was a trip to Europe with his junior squad, where they would compete in France before carrying on to the prestigious Dusika tour in Austria. The trip to Europe would show him what the riders could do and allow him to get to know them as young men. He had met all but one at a training camp in Colorado Springs before leaving for France. The one he hadn't met was a guy from the cornfields of Indiana who people said would be one of the stronger riders in his group.

Chapter 2

THE NEEDLE AND THE DAMAGE DONE

How Greg Strock had looked forward to that spring of 1990 in Europe.

It began with the exchange experience in Spain, and that couldn't have gone better. He settled in with his Spanish family and won three races. So impressed were his hosts that he received the offer to return and ride for Banesto's amateur team. Before leaving Spain to rendezvous with the U.S. junior squad in France, he picked up a heavy cold and a sore throat. After seeing the doctor of the Spanish family with whom he stayed, Strock was put on a course of penicillin. By the time he reached France, he was still a little off-color and not riding well. As it was his first time working under Wenzel, he worried about creating the wrong impression and decided to talk things through with the coach. Wenzel thought the course of antibiotics might be the cause of the rider's poor form.

According to Strock, Wenzel discussed his situation with a French trainer/*soigneur* (carer) whom he had hired for the stay in Brittany. "I didn't know who this Frenchman was," says Strock. "Dark hair, average height, in his thirties, fairly fit-looking, but I didn't get the impression he was a doctor. He didn't do a physical examination or even talk to me, but after he and René had spoken, René came back and said I was to discontinue the antibiotics even though they had been prescribed by a legitimate physician."

The next step, says Strock, went further. "I was told I needed an injection and was given one. As well as that I was given these vials and pills, approximately seven to ten days' worth that were to be taken each day in the case of the vials, and the pills twice a day. They were described to me as 'a variety of pills and extract of cortisone.' Other than a vaccination, this was the first time I had been given an injection. At the time we were starting to find pills pushed into our energy bars. I distinctly remember the first time it happened, biting into a bar and wondering why it tasted so awful. I bit into something strange and could see the cross section of a pill. At first I thought someone at the store must have tampered with it; then I realized our own guys were doing this."

Strock was seventeen at this time, far from home, and concerned about what he was getting into. He asked Wenzel if what he was being given was legal and safe. He was told it was. When he persisted with his questions, he sensed his coach's impatience and got the feeling that what they were doing now was nothing in comparison to what they would be asked to do when they became professional racers. In later years, Strock would look back on these conversations and conclude he and other U.S. junior squad members were in fact being eased into a doping culture. He recalled one particular day from the time in Brittany when he went to his coach's room at the motel and was injected by Wenzel in the buttocks with approximately five to ten cc's of a fluid described by the coach as "extract of cortisone."

After a few races in France, Strock began to feel better, and in the most prestigious race of that spring campaign, the six-day Dusika tour in Austria, he rode strongly to finish eighth. The race ended with a 3.7-mile climb, and halfway to the top Strock attacked to gain a ten-second advantage that, had he been able to maintain it, would have given him overall victory. He was caught seven hundred meters from the finish line but it was, nevertheless, a good effort and one that impressed his coach.

For Wenzel's squad of 1990, the biggest test of the year came at the junior world championships in Cleveland, England. Two days after arriving at their base on July 8, the riders were introduced to a Scottish massage therapist and soigneur, Angus Fraser, who had been contracted by Wenzel to work with the U.S. team for the duration of the championships. Unknown to the racers, Fraser was well versed in providing "medical" care for bike riders, and within the sport he did not have a good reputation. A year later he would be accused of supplying the Australian track rider Martin Vinnicombe with illegal anabolic steroids. Fraser was also accused of writing a letter to the rider advising him on how to use steroids, how to sell them to other riders, and how to conceal the doping from his coaches by saying the steroids were a mixture of legal vitamins.

Strock recalls his first session with Fraser. "René was in the room the first time I got a rub from Angus. He told me to have complete faith in Angus, who he said was a professional and knew what he was doing." Fraser had his own room at the team hotel. As well as massage, his official duties included taking blood tests, supplying medicines and vitamins, and providing injections. He was being paid to do this by the United States Cycling Federation even though he had no medical training. Riders entered his room alone, wearing just a towel, and were asked to lie down on a table. The curtains were drawn and Fraser did his best to make them feel comfortable. "Just relax in here. Don't talk," he told them.

"He wore a blue and white apron," recalls Strock. "Sometimes the injection came first, then the rub; other times the injection came afterward. At one point we were getting two or three injections a day. I questioned René quite a bit. 'What is this?' 'Is it legal?' In his eyes

I was a nuisance. 'Damn it, Greg, if you want to succeed as a pro, you are going to have to learn to trust your trainers and coaches. The pros on the Tour don't waste this kind of energy.'

"In England I was told the injections were vitamins and extract of cortisone. One time we were told it was an ATP [adenosine triphosphate, the energy source used by muscles for short bursts of power] injection. When you break down glucose in the body, you make ATP, and so an injection of ATP was okay. Then for the team time trial, each member of the team was given a caffeine suppository, after which I remember having a horrible stomachache and being curled up in the fetal position. We had a van parked near the start/finish line and it was there we took the suppository. You put it into your anus and it released a stimulant."

The four riders that represented the United States in the team time trial at the championships were Strock, Erich Kaiter, Gerrik Latta, and George Hincapie. Although they were in the silver-medal position for most of the race, the U.S. team's chances were destroyed by a flat tire in the final quarter—a detail quickly forgotten when serious questions were eventually asked about what precisely Wenzel and Fraser were doing with the U.S. junior team in England. A month after the junior world championships ended in England, Strock went to the Washington Trust race in Spokane, Washington. He spoke with Wenzel about how he was feeling, and Wenzel took him to the motel room of fellow USCF coach Chris Carmichael. Strock talked about this incident without actually naming Carmichael, but in Strock's description, this "other coach" had a hard-sided briefcase that he placed on a stand at the end of the bed before opening it. Inside were pills, ampoules, and syringes. Selecting an ampoule and syringe, Carmichael inserted the needle into the ampoule, drew some liquid, and injected Strock in the upper part of the buttocks. Strock says he was told the injection was "extract of cortisone" and that the "other coach" gave the impression of being comfortable with the syringe in his hands. Strock also said he had seen this coach at other races and noticed he often had the hard-sided briefcase with him.

The Spokane race went well for Strock. He was racing against professionals and on the tougher climbs he competed well with them. It was August 1990 and the future stretched out before him, full of opportunity. As well as having that offer to join Banesto's amateur team in Spain at the beginning of the 1991 season, he was also being courted by the fledgling U.S. professional team, Motorola. They told him they were considering taking on three amateurs—Lance Armstrong, Bobby Julich, and him. Armstrong and Julich were a year older and so Strock found Motorola's interest especially heartening. Further encouragement came at the end of 1990 when he was picked for the United States' elite A squad—it was normal for a first-year senior to start with the B squad. Even Armstrong had been originally put on the B squad.

Armstrong and Strock ran into each other occasionally in 1989 and 1990. Though the older Armstrong was coached by Carmichael, they both competed for the U.S. team at the junior world championships in Moscow in 1989, and they met again at training camps in 1990 and in early 1991. "I went to the national training camp in February [1991]," says Strock, "really strong and excited about the future. In the testing we did, I remember my figures were very good. Lance was at that camp and at the time my physiological capacities were higher." An important physiological measure is the athlete's VO_2 Max, the point at which the body can no longer increase the amount of oxygen it uses, despite the intensity of exercise increasing. As a general rule, the higher the VO_2 Max, the greater the athlete's potential in endurance sports. "Pete Van Handl, the guy who did the tests, told me I recorded the highest VO_2 Max he had ever seen from an American cyclist. I was the last rider he ever tested because that evening he got on a plane to fly home and never got there. The plane crashed in Colorado Springs and everyone on board died. In our training races, it was typically Lance and I in a break. We were both riding well."

Armstrong remembers Strock from Moscow. "I raced the road; he was there racing on the track. Smart guy, good guy, blue blood, normal, completely normal guy. But I didn't race with him very much, hardly at all. He raced more with his class, guys like George [Hincapie], Eric Harris, Erich Kaiter, and Kevin Livingston. Guys a year younger than me. Greg was a good bike rider, he wasn't a great bike rider. He was far from that."

In early 1991, Strock left Indiana to begin his year with Banesto in Spain. Soon after arriving, he began to feel sick. The first symptoms were a sore throat and swelling of the lymph nodes. As his condition deteriorated, he developed arthritislike pain in his knees and hips. When he walked up stairs, his knees would suddenly buckle, causing him to stumble and almost fall. Officials at the Banesto team sent him to one of their doctors, who admitted to having no idea what was wrong. Guessing, the Spanish doctor thought Strock had picked up an infection that affected his joints. A course of antibiotics failed to bring about any improvement and suddenly he was sleeping twelve to sixteen hours a day. Unable to come up with the cause of his illness, the Banesto team arranged to fly him back to the United States on April 2.

Back in Indiana, things worsened. Continual tiredness became chronic exhaustion, and struggling to breathe on one occasion, he was taken to the emergency room at the Community Hospital in Anderson. For an eighteen-year-old athlete, the symptoms were highly irregular. One of the first suggestions was lymphatic cancer, an opinion supported by the lymph nodes swelling under both armpits and in the groin. Following minor surgery to remove lymph nodes from the groin, Strock was told the nodes were all matted together, an uncommon distortion. Lymphatic cancer could not be confirmed but neither could it be ruled out.

In June 1991, he went to San Diego for further examination, and from a battery of tests, it was discovered he had human parvovirus B19 infection. At one time or another, this virus affects over 60 percent of the adult population, and in the vast majority of cases, the infection causes relatively minor symptoms: headache, redness of the cheeks, minor inflammation, swelling of the joints. But Strock was severely affected. As well as suffering severe pain in his joints and the swollen lymph nodes, he was truly debilitated. He slept for protracted periods but still felt weak and listless afterward. The good news was that discovery of human parvovirus ruled out the possibility of cancer.

The symptoms lasted through most of 1991 and only as the year ended did Strock begin to think of restarting his cycling career. Chris Carmichael called him and said if he felt up to it, he should try to start training. There was still the possibility of making the U.S. team for the Barcelona Olympics in the summer of '92. But that hope didn't last long. Back on the bike, Strock couldn't summon the strength even to mimic his old form. He got fit and entered races but his body betrayed him. He competed a little in 1992, a little more in 1993, but even on good days, he was some way off his best performances. In his post-parvovirus state, he could not recover quickly enough from the efforts made in a tough race. The struggle went on for more than two years. He even signed for the professional team Saturn and linked up with his old coach Wenzel, but he couldn't regain his athleticism. Then, in that fraught conversation with his coach in Madison, Wisconsin, he finally acknowledged the reality. He was finished.

Strock went back to school and poured new life into what had been a fallback option: he wanted to become a doctor. Indiana University offered him a place in its medical school and he began full-time studies in 1997. On November 26, 1998, Greg Strock chanced upon a discovery that would change the following eight years of his life. "It was a pharmacology class. We were studying steroids, both anabolic steroids and corticosteroids, such as cortisone. And I remember saying, 'Oh, I wonder what the hell is extract of cortisone?' I hadn't really reflected on things from my cycling career but I was suddenly struck by that question. I discovered there was no such thing as extract of cortisone. It is either cortisone or it is not. And I started thinking, these were people with no medical training. Who is to say they were using new syringes? I became pretty angry and I said to my wife, 'I really should try to contact someone about this.' "

Strock also asked himself if the medications he was given as a member of the U.S. junior cycling team could have contributed to the extreme form of parvovirus he contracted. "It is hard to assign definite cause because parvovirus is fairly common," he says. "But the chance of getting the symptoms I got is extremely unlikely. When you see that 60 to 80 percent of the population have had parvovirus but it only affects a tiny minority severely, you do ask questions. Might I have had this virus in my system already and could the administration of cortisone have suppressed my immune system to the point that it allowed the virus to do so much damage?

"I am pretty certain when they said they were giving me extract of cortisone, they were in fact giving me cortisone. That is a lot of cortisone. The biggest question in my mind is 'What were all these other injections?' Will I ever know? I mean, you are dealing with people who would give you antifreeze if they thought it would make you go faster—people with a complete loss of perspective."

In the summer of 2000, Greg Strock filed a civil lawsuit against USA Cycling, René Wenzel, and Angus Fraser. His action was followed by similar suits from his teammates Erich Kaiter and Gerrik Latta. Lawyers for the federation asked for the case to be dismissed because the time since the alleged offenses had exceeded seven years, the cutting-off point in civil cases. Strock argued he discovered the alleged wrongdoing only in 1998 and that he spent the following two years considering whether or not to pursue the case. His wife feared the possible repercussions of a lawsuit, but for Strock the key motivation for bringing the case was the belief that this case should be brought into the open. He felt that what had happened was wrong and that as a future doctor he had a responsibility to expose it. "In general, the American public has a very difficult time trying to imagine the kinds of things that occur as far as elite athletics in this country is concerned. You know that everyone points fingers at the former Eastern Bloc countries and now the Chinese. We can't stand to look at things in our own backyard and we need to," Strock said to

Charles Pelkey in a December 2000 interview with the cycling magazine *VeloNews*.

It is late afternoon on a winter's day in Copenhagen's Town Hall Square. Wearing a black leather jacket and denim jeans, René Wenzel heads for a bar in the square and tells a story that is both defiant and sad. He wants you to know how betrayed he feels by the three riders who filed cases against him, but the animosity he feels for Greg Strock is greater than that which he feels for the others. He and Strock were once such good friends. After receiving that phone call at the Bermuda Grand Prix in 2000 and learning about the lawsuit, Wenzel called Strock but couldn't reach him. He left a message but there was no return call. He tried again, but no luck. Soon he received a phone call from the USA Cycling (USAC) attorney, Bart Enoch, who confirmed Strock was suing both the federation and Wenzel.

Wenzel liked Strock from the start because he was dedicated and got on with the business of making himself a good cyclist. Coaches didn't get trouble from athletes such as Strock, and yet Wenzel knew the young rider wasn't quite the Goody Two-shoes many considered him. If the guys in the squad were having a beer, Strock would be among them; if there were girls who wanted to flirt, he didn't hide himself away. Not that Wenzel disliked him for that; the guy was human, something his coach understood. Among the U.S. juniors he had under his care in 1990 and '91, Wenzel liked Strock, George Hincapie, Fred Rodriguez, and Eric Harris best. Before his marriage to Kendra, Wenzel had a boys' night out that Strock helped to organize.

For his onetime good friend to accuse him of doping him was devastating. "For me, this has been extremely hard. I felt betrayed. I felt lied about. None of the allegations that those guys have come up with are true. None." He then goes through the individual charges. He denies giving any injections, insisting they were given by Fraser. He admits that on one occasion he did get Chris Carmichael to give Strock an injection but that was because the rider asked for it. What troubles Wenzel most is the allegation of doping. "That didn't happen. I believe what came out of Angus Fraser's needles was legal. I had specifically told him I did not want him doing anything that was illegal. Angus was money-happy; he wasn't going to burn bridges with the man who was hiring him. Chris Carmichael gave a vitamin injection, nothing more."

Reminded of the allegation that pills were put into riders' energy bars, Wenzel denies this ever happened and explains how he believes his accusers are confused. "They said I inserted pills into their energy bars, but that did not happen. This is one of the things that is so irritating to me because what I did was tell them about my experiences. I was in France, I experienced pills being put into my food, and I said to them, 'Don't be so stupid to ever do that, don't accept things when you don't know what they are,' and I used the pills in the food as an example. And then it's turned around ten years later."

Though he denies much, Wenzel accepts responsibility for what he now regards as two errors of judgment. "If he [Strock] and his parents and whoever else are upset about the injections, you will not find any protest from me. I have no defense. I agree that people who are not qualified to give injections should not be giving them. I understand that now and I regret that. I wouldn't do it again."

He accepts also that it was wrong to employ someone with Angus Fraser's reputation, but says he did so believing the Scot would not administer illegal drugs. On the two occasions he used Fraser, the appointment was approved by the USCF and the payments made to the soigneur came from the governing body. "In retrospect, I am surprised the federation allowed me to use him."

René Wenzel talks animatedly about the case, and when he mentions Strock, the edge in his voice reminds you he was once a road-race sprinter, a tough guy prepared to fight for his corner. It frustrates him that when it comes to his word against Strock's, he feels people believe the doctor over the cycling coach. The case has taken a toll, virtually ruining his career as a coach, and while he says the controversy didn't cause the breakup of his marriage to Kendra, he says his friends maintain otherwise. They say that since the start of the case, they saw it affect his marriage, and he thinks maybe it did have some effect. Where does he go from here? He still has the house in McKenzie Bridge, Oregon, but after spending most of the last twenty years in the United States, he now thinks about returning to Denmark and picking up the thread of his old life. Most of all, he wants people to understand where he came from. "I know I'm a product of cycling's culture. The injections that some people don't like, I didn't consider them a big deal. The important thing for me was what was inside of the syringes, and I knew it was legal substances. Now I would be concerned about both the injections and the substances. I would like to see doctors kicked out of the sport. Only bike riders who are sick should need doctors."

The difficulty for Wenzel is that it was never simply his word against Strock's. After Strock filed his complaint, Erich Kaiter followed with a second suit and confirmed much of what his former teammate alleged. Kaiter also kept a training diary in which he logged the dates on which he received injections from both Wenzel and Fraser. During the 1990 junior world championships, he detailed a regime that involved forty-two to forty-eight injections over a ten- to twelve-day period. Kaiter also wrote about the colors of the pills he was being given. Toward the end of 1990, Kaiter was laid low by chronic tiredness. Doctors found his white cell count to be extraordinarily high and suspected something was amiss with his immune system. At the age of eighteen, Kaiter's cycling career was over. Nine months later, he was diagnosed with Crohn's disease, a chronic inflammation of the digestive tract that is treatable but not curable.

Two years after Strock's suit, Gerrik Latta filed a third lawsuit against the USCF, Wenzel, and Fraser, in which he backed up the accounts of Strock and Kaiter concerning the 1990 junior world championships in England. The only rider from that four-man time-trial team who did not sue Wenzel was George Hincapie, who has not commented on his experience at the junior world championships.

Wenzel's position was further damaged by a statement from David Joyner, former chairman of the United States Olympic Sports Medicine Committee, on CBS's *60 Minutes*' investigation into the case in July 2001. Joyner said he had learned the USCF had fired Wenzel when they found out what had been going on. This was denied by Wenzel, who said that he lost his job due to the federation's need to cut costs. The USAC backed Wenzel's account, saying he lost his job as part of a restructuring process. Under oath, Joyner said he was told by Lisa Voight, former executive director of the USAC, that Wenzel was fired for doping junior athletes.

A year after Strock's final season in the junior ranks, Wenzel again used Angus Fraser at the Dusika tour in Austria, and after the riders returned from Europe, the USCF received a telephone call from the mother of one of the cyclists, Jimi Killen, complaining about the injections her son had been given. Executive director Jerry Lace expressed his concern to national team director Jiri Mainus, who asked Wenzel to call Killen's mother. Having made that phone call, Wenzel realized that when dealing with minors he had a responsibility to speak with their parents before any injections were administered. This was the late summer of 1991 and many USCF officers and staff, including Wenzel himself, heard the rumors that U.S. junior riders were being doped by their coaches.

On October 31, 1991, Wenzel met for his job evaluation with USCF chief executive Lace, national team director Mainus, and chief financial officer Eileen Johnston. Mainus had three concerns: Fraser's bad reputation in the sport, Wenzel's relationship with Fraser, and Fraser's involvement with the U.S. junior program. During the evaluation, there was stinging criticism of Wenzel, who would later say he felt humiliated by the experience. In his response to the evaluation, written on November 19, Wenzel took issue with his superiors' giving him a "one out of five" on "personal integrity," one being the lowest score. Yet the factual basis for the scathing assessment of Wenzel's performance was never documented, and even though there were many rumors about the team and plenty of reasons for doubting Angus Fraser's integrity, the USCF chose not to conduct an investigation. Neither did they contact any of the athletes, nor their parents, to speak to them about the injections they had received from Fraser.

René Wenzel's contract with the USCF was terminated a year later. The organization's discharge letter, drafted by CFO Johnston on December 10, 1992, cited "restructuring" as the reason for dismissal, though Wenzel himself believed misgivings about Fraser had a lot to do with the decision. At the time, the coach was told he would receive a letter of recommendation from his former employers, but the months passed and there was no letter from the federation. On November 15, 2001—almost ten years after he was let go and over a year after Strock had filed suit against him—René Wenzel received his letter of recommendation from the then CEO, Lisa Voight. In that letter, she highly recommended Wenzel as a coach. Given the complaints filed by Strock, Kaiter, and Latta, the phone call from Jimi Killen's mother, the fact that junior cyclists had been injected without their parents' permission, and the hiring of the unqualified Angus Fraser to "medically" prepare the cyclists, it was strange that Voight should have written that letter, especially since she is currently a member of the UCI's Ethics Commission.

Greg Strock and Erich Kaiter merged their civil complaints against the USCF and Wenzel to form one action, and the case dragged on for six years. The USAC and Wenzel sought a summary judgment against Strock and Kaiter, asserting that their claims were barred by the seven-year statute of limitations and that the two former riders could not prove their illnesses were caused by the medical treatments to which they were subjected. Judge John Kane delivered his verdict on May 5, 2006, and denied the defendants' claim for a summary judgment; he argued that both Strock and Kaiter proved there was a genuine issue of fact about when they should have known the alleged doping caused their illnesses, and because of this, the seven-year statute of limitation ordinarily applicable in civil cases was not relevant. He also found it plausible that repeated cortisone injections had suppressed Strock's immune system and made him vulnerable to an extreme form of parvovirus. He was less convinced by Kaiter's argument that the injections and pills had caused his Crohn's disease and he sided with the defendants, USAC and Wenzel, on that point. What it meant was that USAC and Wenzel lost on three out of four points. Strock and Kaiter were entitled to have their case tried in an open court.

The case was settled out of court in November 2006, six months after Judge Kane's ruling. Strock and Kaiter were each paid $250,000 by USAC and agreed to a confidentiality agreement that prevented them from ever discussing the terms of the settlement. Strock reserved the right to tell the story of his sporting life if ever he was called as a medical witness.

Point thirty-four of Greg Strock's first amended complaint in his case against USAC and Wenzel states, "Medical studies have concluded that this virus [human parvovirus B19] has an 85% correlation with testicular cancer." The complaint was filed in September 2000, four years after Lance Armstrong was diagnosed with testicular cancer. Whether the connection with testicular cancer was innocently made or meant to raise questions about the source of Armstrong's cancer is arguable, but it ensured that the then two-time winner of the Tour de France was mentioned in many of the reports about Strock's lawsuit. Some referred to Strock as a former U.S. teammate of the Texan.

Armstrong was concerned by the implications.

"Strock is a good kid," he said in a subsequent interview. "I called him up because I was getting included in this, and if you want to talk about anything in American cycling, you put my name in it. So they said 'a teammate of Lance Armstrong, same time, same era, same everything.' So I called Greg and I said, 'Listen, you know as well as I do we were never on the same team. You may have been in Moscow racing in the time trial, but I was a year older, racing the road. Completely separate. You need to differentiate there.' "

Committed to a legal battle against USA Cycling and René Wenzel, Strock's hands were full. Last thing he needed was Armstrong on his case. After the phone call, he was happy to clarify the situation, which he did in an interview with *VeloNews.* "Already people are connecting dots between Lance and me, which is inappropriate," Strock told Charles Pelkey. "We were teammates as U.S. juniors and again on the senior A-team in 1991, but we were staggered as juniors, in that he was a year older than me, and I never fell under this coach [Wenzel]."

In the first complaint filed by Strock, Carmichael was named as the "other coach." When that complaint was amended, Carmichael's name was excluded. This change happened because Carmichael paid compensation to Strock to keep his name out of the case. The sum is believed to have been twenty thousand dollars. When asked if he had attended the Washington Trust race at Spokane in 1990, Carmichael said he could not recall if he had. When asked whether he had bought his way out of the Strock case, Carmichael replied that he had been advised by his attorney not to answer that question. René Wenzel tells the story of Spokane very matter-of-factly. "It was an important race and an opportunity for the better juniors, of which Strock was one, to test themselves against older and more highly regarded racers." He says the rider came to the race complaining about not feeling well and Wenzel admits he took Strock to Carmichael's room and asked his colleague if he would give the rider an injection. Wenzel says there is zero doubt in his mind that it was a vitamin injection. "I was present when it was administered."

Strock says he was told the injection contained "extract of cortisone," which he now believes was cortisone. In an interview with the Danish newspaper *Politiken* in November 2006, Wenzel spoke about Carmichael's brief involvement in the Strock case. "Lance Armstrong's personal trainer, Chris Carmichael, was also, initially, involved in the affair, but he disappeared out of the picture quite early on. . . . Carmichael agreed to a settlement very quickly. In hindsight that was probably a smart idea."

Chapter 3

NEW KID, OLD WORLD

At the time the twenty-one-year-old Greg Strock was calling it quits in Wisconsin, the twenty-one-year-old Lance Armstrong was having his first taste of cycling's finest wine, the Tour de France. Armstrong turned up for duty with the Motorola team at the 1993 race blessed with the innocent ambition of a rookie. An exploratory expedition for sure, but not just that. Perhaps no one else expected him to leave his footprints on the race, but he did. And why not? The Europeans didn't faze him. The Tour began on a Saturday afternoon at Le Puy du Fou, a cultural theme park in the Vendée, and if you had been at the official presentation of the teams the night before, watching from above as Armstrong and his eight Motorola mates waited backstage to be called into the limelight, the scene would have told you much. Armstrong tried to act cool but that was impossible. Out front a band played, thousands of people clapped and cheered, television cameras were trained on the stage, a big screen relayed pictures, and the kid was in his element. "This," he whispered to his Australian teammate Phil Anderson, "is what bike racing is all about. This is what life's about."

By then Anderson was in the twilight of a successful career, at the point where people were asking him about the old days, but seniority and performance made him one of the big men of the Motorola team. He had competed against Hinault, LeMond, Kelly, Moser, Saronni, Fignon—the patrons of the eighties. And even though Anderson had seen a lot, Armstrong interested him. You can tell when a kid has something, and he had seen it in Lance—ability, power, attitude. Especially attitude. You raced against guys to beat them, not to show them respect. Perhaps Anderson saw shades of his early self, when he, too, used to ruffle the feathers of the big chiefs. In races, Armstrong bugged the hell out of you: "Can I attack now, can I go now?" Anderson would tell him to wait, to calm down, too far out, just get back in the pack. Ten minutes later he was there again. "Can I go now?" He chipped away until you could take no more. "Yeah, okay, go. Don't look back."

You would have seen the kid's zest for life at the *Village Depart* in the town of Avranches on the sixth day of the 1993 Tour. The *Village* is temporary and tented, erected to allow the Tour's corporate friends and their invitees to schmooze with riders. That morning Armstrong maneuvered his bike slowly through the forest of people, past the coffee counters, the trays of salami, the rows of fresh peaches, the reporters with their notebooks, ordinary fans with their cameras, and because he was young and new, he smiled on all who smiled. An Italian journalist asked who he thought would win that day's race. "Sciandri," he said, "my teammate Max Sciandri." Two pretty girls walked past, pretending not to notice him; his eyes followed them through the crowd until he could see no more. His feet now on the ground, his hands resting on the handlebars, he stood there like the kid he was—youngest rider in the race, a child at his first circus. Yet he wasn't there just to watch. Before he left, he wanted people to know him: *Remember what Lance did at his first*

Tour de France? As he stood there, the two pretty girls returned his way. "*Monsieur Armstrong, s'il vous plaît,*" one of them said as she passed him a pen and paper. How pleased he was to see them again, to write his name with the L so big he could fit the rest on the horizontal line. He tried to detain them, but with Gallic hauteur, they went on their way.

This was Armstrong's first full season as a pro and it wasn't just Anderson who saw the potential. Every guy on the team could see he was different. What kind of rookie comes to Europe with a business card and baseball caps with a personalized logo? Before the marketing world knew he existed, Armstrong was developing the brand. This went back some way. When he was a kid, starting out in a triathlon, he trawled local businesses in Austin for sponsorship, and when everyone said no, he took his tank top to a printer and emblazoned it with the name of his greatest sponsor: I LOVE MY MOM, it said. And some guys just talk, but he could compete. "Holy shit," the Motorola guys said when they heard the result from the 1992 First Union Grand Prix in Atlanta. Riding for the U.S. national team, Armstrong outsprinted the U.S. national champion, Greg Oravetz. Everyone knew Oravetz—superstrong, a beast who didn't lose to amateurs. Until that day in Atlanta.

Armstrong joined Motorola after the Barcelona Olympics in the summer of 1992, and even if he finished last in his first race as a professional, the San Sebastian Classic, they noticed the grit he showed in refusing to quit. A couple of weeks later he was second in the Championship of Zurich, a reputable end-of-season classic, and those Motorola teammates who weren't in Zurich can recall where they were when they heard the rookie got second. New pros didn't do that kind of thing. And of course his teammates noticed his attitude. They saw little in his character that spoke of humility and compromise. And while there was much he didn't understand about Europe, they learned it was best to let him come and ask. Tell him what he should do and the conversation often came to an abrupt end.

There were the fun sides, too. He would joke around, talk big without bragging, and when the mood took him he could be charismatic. In that first full season in Europe, 1993, he hooked up with Frankie Andreu, who had been with the Motorola team from the start and was into his fifth season as a professional. Andreu was living in Belgium, and when Armstrong suggested they move to Como in Italy, there was no objection from Andreu. Motorola's team doctor, Max Testa, lived in Brunate, just outside Como, fellow American riders Bob Roll, Jeff Pierce, and Ron Kiefel had all lived there, and teammate Andy Hampsten was just three miles outside of the town. Max Testa found them somewhere to live, a two-bedroom place in an apartment block that overlooked a supermarket. Their first training ride was a five-hour spin around the lake at Como, remembered by Andreu because Armstrong accelerated away from him on the way home, left him muttering, "You son of a bitch." Afterward, they ate watermelon on the balcony and shot the breeze about their form and what they were going to cook for dinner that evening.

After training with Armstrong, teammates came back with stories. Motorists who honked or cut in too tight always got an Armstrong reaction—a yelled profanity, a one-finger salute, or every so often the driver stopped and a little hell broke loose. They remember the time he lived on Cap Ferrat in the south of France: some guy buzzed him, Armstrong reacted in the usual way, and suddenly he was standing face-to-face with his neighbor's gardener—a situation made more interesting by the pair of shears in the gardener's hands. Another time in Italy a guy got out of his car with a huge wrench, but the thing about Armstrong was that it didn't matter—shears, wrench, whatever, he didn't back down. "I remember one training ride with Lance," said Frankie Andreu. "We were in our Motorola jerseys and as a car moved past us, this guy shouted, 'Hampsten! Hampsten!' mistaking Lance for Andy. Lance flipped. 'Agh, agh, Armstrong! Armstrong! I'm Armstrong.' And he had that mentality: 'I'm Armstrong, I'm the leader. How dare you not know me? You see this jersey—it's Armstrong. Nobody else.' Andy was a leader, but to Lance he was nothing. 'I'm the guy.' "

Arrogance was tolerated because it was part of what made him a good racer; people said it was the natural exuberance of a talented young rider. When the team needed a win, there was a good chance the kid would deliver it. Motorola went into the 1993 Tour de France with Hampsten as leader and well capable of finishing in the top ten, possibly even the top five. But the team also needed to do something along the way—a daring exploit, a dramatic victory, a moment that put the team into headlines and reminded Europeans of Motorola mobile telephones. Looking out for his teammate, Andreu scanned the Tour route for a stage that might suit Armstrong. Four or five days into the race, he found it—the eighth leg from Châlons-sur-Marne to Verdun.

"Lance," he said, "I've got one for you. I've got a stage for you."

"The one into Verdun?"

"Yeah. That's the one for you."

"I know it, I know it."

By the time the race reached Châlons-sur-Marne, the evening before the leg to Verdun, Motorola's need was getting close to desperate. That very day they got three guys into the seven-man breakaway that dominated the end of the race. Three blue jerseys in a group of seven; the only way Anderson, Sciandri, and Alvaro Mejia could lose was by messing up. Back in the peloton, Armstrong high-fived his teammates—the eggs weren't just counted, they were hatched. But from certain victory, the Motorolas plucked only second place. That evening, not much was said as Armstrong's brooding disapproval discouraged the lighthearted banter that comes with dinner. It wasn't the moment for Andreu to remind Armstrong he could win into Verdun, but they both knew it was a route made for him.

Ten miles from the finish into Verdun, there was the tough 1.2-mile Côte de Douaumont climb that would eliminate the sprinters but that wasn't long enough to give the climbers a decisive advantage. Somewhere in the great expanse between the mountain goats and the rapid finishers, Armstrong could be found. He had the explosive power to survive on short, sharp climbs, and it would take a strong man to beat him at the finish. If the race went as Andreu had foreseen it, the hill would eliminate the pure sprinters. At the foot of the Côte de Douaumont, the Italian Claudio Chiappucci was one of the first to accelerate, the then two-time Tour winner Miguel Indurain went after him, and so, too, did Gianni Bugno. After him went the powerful Swiss, Tony Rominger. Chiappucci surged again, and the Frenchman Laurent Brochard reacted, as did Armstrong. The two pursuers caught Chiappucci and sheltered in his slipstream to draw breath.

Attack, counterattack, each rider hoping he would draw the long straw and end up in the break that went clear, because eventually one of them would. Being involved in a game of chance didn't appeal to Armstrong and he tried to cover every move, so that when the decisive split came he would be in the right position. His lungs burned from the effort, and when three riders escaped just after the summit, he was perhaps one hundred yards adrift and wasted from the sustained effort. He wanted it to all come together again while he recovered, but one glance behind him suggested it wouldn't. Two guys were coming up behind him—after that a widening gap back to the pack. So Armstrong kept going, was joined by the two from behind, and the three rode together to join up with the three in front. The six-man breakaway from which the day's winner would come had formed.

Armstrong x-rayed his five companions for levels of threat. Only the French rider Dominique Arnould worried him, and as the riders neared Verdun and jockeyed for position, Armstrong raced directly behind Arnould. The French rider didn't want Armstrong breathing down his neck, so he slowed, feigned an attack, all the while trying to get his rival to take the initiative. Armstrong was nerveless. Instead, he baited his rival. "Hey, man, you're on a French team, you're a French rider, this is the Tour de France. You're the one who can't afford to mess up." Arnould might not have understood the English but he got the message. As it turned out, another French rider, Ronan Pensec, was the bigger threat. He surged early and gained a decisive advantage. Arnould hadn't the speed to peg him back, and that left Armstrong, who had. Soon the gap was closing—Pensec sensed his rival coming between him and the crash barrier—and he instinctively veered a little to his right to narrow the gap.

"We were right at the hundred-and-fifty-meter sign," said Armstrong afterward, "so he can't come over, that's against the rules, but he kept moving toward me and I didn't know what I was gonna do. I decided to just yell as loud as possible and maybe freak him out a little. I mean, no matter how bad an ass you are, you are still going to get a little hesitant when somebody screams. And so he hesitated and that was my gap. Right there, that's all I had."

He went past Pensec so fast he had time to raise both arms in celebration before crossing the line; the Mexican Raul Alcala got past Pensec to take second. Armstrong afterward tried to explain what got into him at that vital adrenaline-driven moment. "Physically I'm not any more gifted than anybody else, but it's just this desire, just this rage. I'm on the bike and I go into a rage, when I just shriek for about five seconds. I shake like mad and my eyes kinda bulge out. I swear, I sweat a little more and the heart rate goes, like, two hundred a minute . . . and that's heart man. That's not physical, that's not legs, that's not lungs. That's heart. That's soul. That's just guts."

The dramatic victory at Verdun foretold great days to come, and Jim Ochowicz, the team director, showed maturity in his management of the rookie. That is, he indulged him. *What do you think, Lance?* Ochowicz agreed he should ride the first two Alpine stages that came in the middle of the race and then pack his suitcases and catch a ride back to Como. The plan was for the kid to learn about the Tour but to leave before the lessons became painful. Phil Anderson stayed by his side for those two mountain stages, and even if the primary objective was to gain experience, the time losses were still considerable. Armstrong finished twenty-one minutes down at the end of the first day, then a further twenty-eight minutes down the second day. Anderson was vastly experienced, and he sensed the kind of racer Armstrong would be. "He was a one-day rider. I thought he could never, ever, win the Tour de France. Even he wouldn't have thought he could have won the Tour. He couldn't climb and he couldn't time-trial, two things you have to do to win the Tour."

After Armstrong left the race, Motorola's tour got even better. The Colombian Alvaro Mejia and Hampsten rode well to stay close to the leaders almost every day in the mountains. Miguel Indurain and Tony Rominger were by some distance the strongest in the race, but after that, the two Motorola riders were as good as any of the others. They finished fourth and eighth overall and Mejia's fourth was an outstanding achievement for a rider hired for a song and expected to be nothing more than another member of the choir—that is, a support rider in the mountains. Fears that the sponsors might leave at the end of the 1993 season were banished by the Tour performance. Then, a little more than a month later, Armstrong won the world championship road race in Oslo with a ride of almost unimaginable power. For a twenty-one-year-old to dominate the world championship road race, as Armstrong had, was an extraordinary feat. The promise that he would become one of the sport's best one-day racers had been realized in his very first season.

The future for him and the team could hardly have looked better.

If you think of the sport of cycling as a great oak tree, European cycling is the trunk. For it is on the old continent that the sport began. On small roads in France, Belgium, and Italy, professional cycling constructed its DNA. Those black-and-white photographs of dusty cyclists in woolen jerseys, a spare tire looped around their torsos, pain etched on their faces, were the images that defined the sport. The suffering was an offering, a gift freely given by sportsmen to their public. They drove themselves to God knows what lengths, endured the most terrible conditions, because at the end of their day, we would think more of them and more of ourselves. The severity of the sport examined a man's character, illuminating his nobility but also addressing his baser instincts. Perhaps most of all, it offered a challenge that allowed man to transcend his everyday self. Standing on the side of the road, we shook our heads in admiration: *Did you see that? The wasted body? The picture of suffering? The eyes revolving in their sockets?*

In the early days, the sport attracted diverse characters—many of them men with not a lot to lose—who dealt with the demands in their different ways. Refuge was sought wherever it could be found: pills to take away the pain, drinks that brought temporary feelings of well-being. And who could blame the wretches for doing what they could? There were no doping laws then, nothing that made it wrong to hitch your wagon to an outboard motor. That the stimulants or the alcohol or any other poison might put a man's health at risk was not a deterrent; in fact, there was a certain appreciation for the recklessness of those prepared to go so far. As the sport developed, so, too, did the doping culture and the participants' sense of entitlement. Doping became *their* business. Their value system was unique: it was wrong to speak in a way that harmed cycling's image, but there was almost total indifference to conventional morality. Riders accepted money to pull ever so slightly on the brakes and ease a rival's path to victory, because the guarantee of payment was better than a shot at it. As for that fleeting moment of glory, well, that didn't feed a man's family.

For those coming from the outside, this sporting culture took some getting used to. If you were a blond-haired, blue-eyed Californian like Greg LeMond, it was a bit of a shock. LeMond went to the world championship road race at Barcelona in 1984 thinking everyone was there to win. "The Italian Moreno Argentin was riding in my slipstream all through the race, just making sure he followed every move I made. I said, 'What are you doing?' Eventually he said he would help me if I agreed to pay him ten thousand dollars. I was outraged and thought, 'Damn you.' " After the race LeMond publicly complained about the Italian's readiness to prostitute his talent. For speaking plainly about the sport's darker side, LeMond was made to feel like the guy who spoiled the dinner party by telling a distasteful joke. Had LeMond been born in Liège and not Lakewood, perhaps he would have accepted the idea that you don't *crache dans la soupe* (spit in the soup), because by pointing out the sport's tawdry side, the accuser hurt the sport's image.

When 7-Eleven became the first U.S. team to compete regularly on the European pro circuit in the mid-1980s, the American riders on the team crossed the Atlantic with definite ideas about how they would go about their business. When they were accused of not being able to handle their bikes in the heaving mass of a bunch sprint or in the brutal crosswinds of northern France and Belgium, they would not stand for it. Instead they responded with exaggerated aggression, snarled when Europeans tried to belittle them, and refused to be cowed. Doping was the other big issue. "We were going to Europe," says Andy Hampsten, "and we wanted to do these big races, but we were going to do them being the men we were and not take drugs to get ourselves to an artificial level. That might sound high and mighty, but for me—and I think I speak for just about everyone on the team—it was our strength. We could see how the drugs of that era worked. Guys who would beat us in the spring faltered in the summer, when the drugs didn't seem to work so well."

In the 1980s, amphetamines, testosterone, and corticosteroids were widely abused by cyclists. As they were easily detectable at doping control, amphetamines were used only in races with no medical controls and for training purposes. Testosterone was difficult to detect, corticosteroids were undetectable, and both were widely used, in and out of competition. These drugs accelerated recovery and helped to mask pain. Cortisone and other corticosteroids affected the body's water and salt balance, and it was widely believed they weren't as effective in the heat as they were in cooler spring or fall weather. While they gave an advantage to those who used them, they weren't so powerful as to leave clean riders without hope, and neither were they strong enough to transform an average rider into a consistently successful one.

Though it was treason to speak publicly about the doping culture, drugs were part of cyclists' lives. Signs of cortisone abuse were easy to pick up, and riders looked at one another in the manner of young ladies at a health club. "We would look at the red and pumped-up faces of some of the European guys and not know what to think," said Frankie Andreu. In those days you could tell from the bloated faces, from the power of the guy with the big, almost fat legs; it was discernible even in the superficial gash from last week's crash that didn't heal because cortisone suppressed the body's immune system, lessening its ability to heal. At all but the biggest races, drug testing was lax and teams were told in advance at which events their riders would be tested. Where there were no controls, there were plenty of amphetamines. And in a sport where doping was an entitlement, few insiders were going to stand up and speak against it. Those who did were shouted down.

And, relatively speaking, those were the good days.

Chapter 4

THE TERRIBLE ELIXIR

In the 1980s it was possible for a young and talented rider to come from the United States and make his way in Europe without doping, as evidenced by the success of Greg LeMond in the middle of the decade. Young, fresh-faced, richly talented, LeMond competed successfully against the cortisone/testosterone generation. He finished third in his debut Tour de France in 1984, second the following year, and then won it in 1986. Bernard Tapie, owner of the La Vie Claire team, for whom LeMond rode in '85 and '86, said LeMond was one of a very small number of riders he believed was clean. Another who would have been included in that select group was a tall young American with a bright smile and an altar boy's manners.

Andy Hampsten was born in Columbus, Ohio, and raised in North Dakota. He once joked that if you've seen the Coen brothers' splendid film *Fargo,* you'll know why as a teenager in North Dakota he faced his bike south and started pedaling. In the sport of professional cycling, he was one of the guys you trusted. He spoke out against doping, delivered the kind of even performances that no one could question, and was the same rider at the end of his career that he had been at the beginning: a fine climber and great athlete who didn't measure up against the best in the race against the clock. In the early summer of '85, Hampsten was riding his first pro season with the Levi's-Raleigh squad in the United States when the rival 7-Eleven team offered him the opportunity to ride Italy's national tour, the Giro d'Italia. At the time he had less than four months' experience as a pro; he had never competed against the best European pros and had never ridden a grand tour.

Almost three weeks into the Giro, 7-Eleven's directeur sportif, Mike Neel, still believed Hampsten could achieve something. He picked out the twentieth stage—a short, explosive race to the summit of Gran Paradiso in the Italian Alps. Just twenty-five miles through the valley and then an eleven-mile climb to the finish, it was a race for climbers, and Hampsten could soar with the best of them. As the stage did not begin until midafternoon, he reconnoitered the climb that morning with teammate Jonathan Boyer, with Neel driving the team car behind them. From the top, all three traveled down the mountain in the car and in his eternally optimistic way, Neel kept telling Hampsten he could win. The other climbers wouldn't give him much latitude, Neel warned, but if he attacked early, where the climb was most difficult, there was a chance. "Right here," said Neel, indicating a stretch of road near the bottom, "this is where you've gotta go." Often, the key to dealing with Hampsten was to steal inside his head, plant a positive thought, and wait for it to blossom.

That afternoon, Neel had the team primed and they were all ready to sacrifice themselves. Hampsten was so taken by the plan that he wore a one-piece skin suit normally reserved for time trials, a choice of clothing that drew plenty of European smirks. Hampsten didn't care: just thirty-six miles, he wasn't going to need anything to eat, there was no descent, he wasn't going to get cold. But the expectation did weigh on his twenty-three-year-old shoulders, especially when he remembered the opposition. Bernard Hinault, LeMond, Giuseppe Saronni, and Francesco Moser were in the race and all were world champions. But in the end, he could worry all he wished—what he couldn't do was back out.

In the jockeying for position on the approach to the first ramps of the climb, the Italian Roberto Pagnin attacked. Perfectly positioned near the front, Hampsten bided his time as the peloton's gradually increasing tempo ate into Pagnin's advantage. The breakaway was on the right side of the road, and as the Italian was overhauled, it seemed as though he were being sucked back into the steaming pack. Then a small hole opened at the front of the peloton and he disappeared into it. There was a temporary respite as people waited for the next move. "Where are you? Where are you?" shouted 7-Eleven's Bob Roll as he flicked his head sideways. "I'm here, right behind you," said Hampsten. And then, right on cue, Hampsten bolted. The trick for the rider attempting to escape is to sustain an all out effort for as long as possible, not look back until there is nothing to see. Hampsten kept going for what seemed like an age, but it wasn't much more than two or three minutes, perhaps less. His backward glance caught only the Spaniard Marino Lejarreta and he was too close for Hampsten's comfort. At that moment and throughout his career, Hampsten heard mocking inner voices: *You ain't going to make it.* But even back then he knew how to deal with doubt. "I don't *have* to win," he told himself. "All I *have* to do is go as hard as I can. The world isn't going to fall apart if I don't win. I want to win but I don't have to," and so on. That was the way he calmed himself, and in that state he rode his best.

Halfway up that mountain, a small blackboard carried by one of the motorcycle outriders told Hampsten his lead was twenty seconds. He did the math—five miles, twenty seconds; the other riders had to find four seconds per mile. That was easy for a pack with the scent of blood in its nostrils, so he concentrated on staying relaxed, letting his legs do the work. His lead increased to thirty seconds, then forty, and then there were fewer than two miles to go. At that point, or perhaps a little closer to the finish, he redid the math, this time including the possibility of a fall costing him twenty seconds—and even with that, they still couldn't catch him. Then he felt euphoria flowing through his veins, swelling his heart, bringing tears to his eyes, and, magically, his legs turned effortlessly. You can take whatever drug you like, performance-enhancing or recreational, and it couldn't touch that feeling. Three years later Hampsten would win the Giro d'Italia—not just one leg but the whole race. He would twice win the Tour de Suisse, the Tour of Romandie, even the treasured Alpe d'Huez stage in the Tour de France, but nothing ever surpassed what he felt on the last pull to the summit of Gran Paradiso.

That was 1985, a year when it was possible for a young, inexperienced, and clean racer to achieve a victory as truly heroic as that.

Ten years later, the sport had undergone terrible change, transformed by the rampant abuse of a new drug, erythropoietin. Produced by the kidneys, the natural hormone erythropoietin regulates red cell production. Its synthetic equivalent—recombinant erythropoietin (r-EPO)—first appeared on the market in the late 1980s. For the medical world, this was a hugely important drug used in the treatment of anemia resulting from kidney failure or from cancer chemotherapy. Not only was the new drug effective in reestablishing a red cell population, it led to few side effects. In a short time, r-EPO was acknowledged as a wonder of medical science. In 1989, Les Earnest, a director of the United States Cycling Federation (USCF), wrote that r-EPO had the potential to help athletic performance and to bring an end to potentially more hazardous blood transfusions. Before suggesting r-EPO should be used by sportsmen, Earnest wrote to Jerry Lace, USCF executive director, asking that a study be done to determine if the drug could be used safely. Earnest was reassured such a study was under way, which it was not. Earnest was not an advocate of doping and had been critical of the blood doping program undertaken by members of the U.S. cycling team at the 1984 Los Angeles Olympic Games. Though it was not illegal at that time, blood transfusions were unethical. Even so, U.S. riders used them for the '84 Games with the support of the then national coach Eddie Borysewicz, three members of the medical team, and a high-ranking official.

Earnest would later admit his first reaction to r-EPO was naïve, but even to the morally scrupulous the drug was tempting. It wasn't banned; it didn't involve the drawing, storage, treatment, and reinfusion of blood—all it did was replicate the effects of altitude training. To the morally unscrupulous, it was an elixir. By artificially elevating the athlete's red cell count, the drug greatly enhanced his oxygen-carrying capacity, and in endurance sports, that makes everything much easier. Peak performance can be sustained for a longer period and the toll on the body is less than it would be for those with lower oxygen capacity. In a three-week race such as the Tour de France, the use of r-EPO allows the rider to maintain a level of freshness and physical well-being that is simply not naturally possible. Endurance sports had never known a drug like this.

We cannot know for certain where r-EPO was first abused by athletes, but the suspicion has always been Holland, where a number of cyclists died suddenly in the late 1980s and early '90s. Death generally resulted from a heart attack suffered while the rider slept, and even those with a rudimentary understanding of how r-EPO worked figured that as the drug produced more red cells, the blood thickened and caused clots when the flow slowed during sleep. After the clot came fatal cardiac arrest. Bert Oosterbosch and Johannes Draaijer were the two best-known Dutch riders to have died at this time—in 1989 and 1990 respectively—but there were more. The tragedies inspired no official inquiry and the cause of death was not proven.

But the deaths sent shivers of alarm through the peloton, and riders sought medical advice on how to use r-EPO without endangering their lives. Professor Francesco Conconi, rector at the University of Ferrara in northern Italy, was regarded as an authority on blood doping and he was much in demand by both the authorities who needed a test to detect r-EPO and those racers who wanted to abuse it. Conconi worked with a group of younger doctors at his university in Ferrara who were also well versed in the effects of blood doping and, it would later emerge, other doping products. By the early '90s, Conconi and his associates were supervising the medical preparation of their country's best cyclists, and in no time, Italy became the dominant country in world cycling.

In addition to the appalling human cost, r-EPO had catastrophic consequences for sporting competition. If used correctly, an athlete's oxygen-carrying capacity could be increased by 20 percent or more. That meant a well-trained elite athlete using r-EPO in an endurance event was virtually certain to beat a clean rival, and in no sport was the drug as influential as in professional cycling. By 1994 it was clear many Italian teams were using r-EPO. Equally pressing was the need for a vigorous response from the anti-doping authorities. At this time, the anti-doping movement was disorganized and uncommitted. Cycling's governing body, the Union Cycliste Internationale (UCI), talked about the number of drug tests it conducted and the relatively low number of positives, disingenuously jumping from there to the conclusion that the sport didn't have a problem. In doing so, the sport's governing body ignored the fact that two of the most popular and most potent drugs—r-EPO and human growth hormone—were undetectable.

Perhaps the most eloquent demonstration of cycling's attitude toward doping came in the collective response to Paul Kimmage's eye-opening book on the sport's sickness, *Rough Ride*, first published in 1990. Kimmage had been a professional for four seasons (1986–89) and depicted a sport with a serious drug problem about which the authorities and competitors were in denial. He described races with no controls, riders injecting themselves during races, collusion between race organizers and teams to facilitate doping, pervasive doping on most teams, and a conviction among riders that if you didn't dope, you raced at a disadvantage. Riders and officials vilified Kimmage for writing what they saw as an "anti-cycling book," and, in a time-honored cycling tradition, they dismissed him as just a small rider who'd never won much. That was easier than facing the truth. What might now seem remarkable was the UCI's indifference to the allegations; but, as much as the riders and team officials, the parent body was complicit in the tolerance of doping.

Just how bad things had been was shown by Sandro Donati's experience with sporting authorities in Italy. Donati was appointed coach to his country's middle-distance runners in 1981, and shortly after starting the job, he encountered Professor Francesco Conconi at a scientific conference. In addition to being head of the University of Ferrara in northern Italy, Conconi was a prominent member of the Italian Olympic Committee (CONI), and he congratulated Donati on becoming a national coach and told him about the program just begun with the country's elite athletes. Basing their model on that used with Finnish athletes in the '70s, they were going to draw blood from each athlete, refrigerate it, enrich it, and transfuse it two or three days before an important competition. According to Conconi, this was sanctioned by both the Italian athletics federation and CONI, and the transfusions would improve a fifteen-hundred-meter runner by three to five seconds, a five-thousand-meter runner by fifteen to twenty seconds, and a ten-thousand-meter runner by thirty to forty seconds. He asked Donati to nominate his most suitable eight-hundred-meter and fifteen-hundred-meter athletes for the program.

Donati was appalled by the proposal but not immediately sure what to do. Blood transfusions were not illegal, and if the athletes wanted them, there wasn't much the coach could do. Gathering seven of his best middle-distance runners, Donati told them of Conconi's blood doping program but added he wasn't in favor of them joining. Not one of the athletes wanted to do it. Pressure to comply was exerted by the Italian athletics federation, and Conconi invited two of the runners to his university in Ferrara to see firsthand what he had in mind. Still the athletes refused to submit. Federation officials were displeased with Donati, who they felt turned the athletes against the transfusions. The coach was told he would lose his job directly after the 1984 Olympic Games in Los Angeles. It didn't help Donati that other Italian coaches and their athletes participated in the program. Donati wasn't sure to whom he should complain. After all, CONI, for whom he worked, approved the program. Eventually, Donati used his contacts inside the Italian parliament to get senior politicians to take an interest. They decided blood transfusions for performance enhancement was doping and they outlawed the practice. Soon afterward, the International Olympic Committee (IOC) declared blood transfusions to be a method of doping and put them on their banned list. In the war against doping, Donati had just won a battle, but in the wider sense not much had changed. Professor Conconi's standing in the official world of sport was reflected in his continued membership on the medical commissions of both the IOC and the UCI.

How out of touch the authorities had become was apparent from their reaction to the cycling deaths of the early '90s and to the widespread evidence that r-EPO was distorting sporting competition, especially in cycling. Realizing they needed a test to identify r-EPO, the IOC funded Professor Conconi to do the research. Convinced Conconi was not the man to combat this problem, Sandro Donati began his own investigation into doping in Italian cycling, and by guaranteeing anonymity to twelve key figures in the sport, he unearthed much important information. Namely, he learned anti-doping tests were not working because so many banned products were undetectable. He also discovered that abuse of r-EPO was rampant and that the man chosen by the IOC Medical Commission to come up with a solution had actually negotiated a deal with one of the top cycling teams, Carrera, to administer r-EPO to their riders.

Officially, Conconi was working to find a test to detect r-EPO in urine. He reported to the head of the IOC Medical Commission, Prince Alexandre de Merode, and there were public statements about how close the Italian professor was to a definitive test. Conconi claimed he was working with a sample group of twenty-three amateurs, and by administering r-EPO to them, he was able to measure the effects on their blood values. From this research, he was supposedly on the brink of a breakthrough. The r-EPO used by Conconi was supplied by a company in Mannheim, Germany, following a written request from the IOC to the company. Conconi administered it to the athletes over a forty-five-day period, but his research was not based on amateur athletes. Rather, it was based on a group of twenty-two elite athletes, mostly cyclists from the Italian team, Carrera. The University of Ferrara professor was simultaneously being funded by those trying to catch the cheats and being paid for r-EPO treatments by the cheats themselves.

According to a subsequent police investigation, the twenty-third athlete in Conconi's r-EPO study was listed under the code name "Ferroni" and was in fact the professor himself. Pierguido Soprani, the prosecutor in the case against Conconi, claimed the professor, who was fifty-nine at the time, recorded a hematocrit value of fifty-seven on September 3, 1994. Hematocrit is the amount of red cells in the blood expressed as a percentage of total blood volume, and the average value for a male is in the midforties. The Italian cycling journalist and anti-doping campaigner Eugenio Capodacqua didn't need to see the police file to believe Conconi was on r-EPO. In September 1993, Capodacqua holidayed in northern Italy, but that was just an excuse to allow him to compete in a time-trial race to the summit of the 16.9-mile Stelvio Pass in the eastern Alps. Capodacqua traveled with his friend Alfredo Camaponeschi, who had been a good amateur cyclist and was still a very fit thirty-four-year-old. Capodacqua hoped to do a good time, and his friend expected to. The journalist rode well, setting off four minutes after Hein Verbruggen, president of UCI, and overtaking him halfway to the top. Camaponeschi rode very strongly, but afterward, he and Capodacqua talked about nothing except the performance of a man in his late fifties. Professor Francesco Conconi climbed the Stelvio fourteen minutes quicker than Camaponeschi and an unmentionable amount of time quicker than Capodacqua. The journalist and his friend thought this wasn't natural, an observation proven by the subsequent police investigation.

For a long time Conconi's connections helped to protect him. At different times he has sat on committees of UCI, IOC, and CONI. He is a longtime friend and cycling partner of Romano Prodi, the former Italian prime minister, and at the time he was doping his "sample" athletes, Conconi enjoyed a close friendship with Prince Alexandre de Merode, then chairman of the IOC Medical Commission. So close were they that in January 1992, Conconi invited De Merode to the University at Ferrara and awarded him an honorary doctorate in medicine and surgery. According to Conconi, the award was recognition for De Merode's work in the fight against drug abuse in sports.

When Sandro Donati tried to point out that Conconi was a fraud, officials in Italy considered him a maverick who should choose his words more carefully. Donati, though, was undeterred. After concluding his 1994 investigation into cycling's drug problem, he wrote a fourteen-page report that he called "The EPO Dossier" and gave it to the president of CONI, Mario Pescante. It elicited no response from the president and the only thing Donati noticed was a new coolness in the president's attitude toward him. Raffaele Pagnozzi, the general secretary of CONI and to whom the report was also sent, gently told Donati he wasn't helping himself in CONI by devoting so much time to doping.

Two years later two journalists from the Italian sports paper *La Gazzetta dello Sport* asked Donati for help on a doping investigation and were referred to Dr. Flavio Alessandri, one of the original sources for "The EPO Dossier." Alessandri told the journalists what he knew and mentioned he had revealed all this information over two and a half years before. When the two journalists went back to Donati, he said the report had been given to the president of CONI. At first, Mario Pescante denied its existence but then admitted he had received it. He would eventually be forced to resign as president of CONI following an investigation into the covering up of positive drug tests at the IOC-approved laboratory in Rome.

The police investigation into Francesco Conconi and his two colleagues, Ilario Casoni and Giovanni Grazzi, lasted five years and ended when they were acquitted on charges of committing sporting fraud. Central to the case against the three doctors were the medical records of a number of top cyclists in the early 1990s and blood tests revealing substantial fluctuations in their hematocrit levels. Franca Oliva, the presiding judge, considered that the fluctuations were indicative of r-EPO abuse, but on their own, hematocrit levels are not conclusive proof of r-EPO. But it was a strange kind of acquittal. In her summary, the judge called the doctors morally bankrupt and wrote a damning forty-four-page report.

> The accused [Conconi, Casoni, Grazzi] have, for several years and with systematic continuity, aided and abetted the athletes named in the court indictment in their consumption of erythropoietin, supporting them and de facto encouraging them in that consumption with a reassuring series of checks on the state of their health, with examinations, analysis and tests designed to assess and maximise the impact of that consumption with regard to sports performances. Therefore, on a point of law, the crime as originally charged against the defendants still stands.

Conconi and his associates did not appeal Oliva's judgment.

Imagine the mood on the Motorola team bus on the evening of Wednesday, April 20, 1994. It was parked in the old Belgian town of Huy, nineteen miles southwest of Liège, and the bunch of riders who flopped and spread themselves on the seats were tired and disgruntled. They had just spent five hours racing the Flèche Wallonne classic and had nothing to show for their efforts. Three days before, Lance Armstrong was second in the Liège-Bastogne-Liège classic, and though that wouldn't have pleased a man who'd set out to grind every sporting rival to fine dust, the rest of the team knew second was a lot better than nothing. Flèche Wallonne was their second chance in the Ardennes, but it had now slipped past and not one of their guys got into the top ten. They just knew their team was better than that.

As a group, they didn't know what precisely was happening in their sport, but they sure as hell knew something was going on. At Liège-Bastogne-Liège, where the team worked hard for Armstrong, the race was won by the Russian Eugeni Berzin, riding for the Italian

Gewiss team. Berzin won on his own, one minute and thirty-seven seconds ahead of Armstrong, who beat another Gewiss rider, Giorgio Furlan, in the sprint for second place. If first and third was a good day's work for the Gewiss team, it paled in comparison with the team's performance at Flèche Wallonne. It would have been normal for Berzin and Furlan to have been exhausted after their seven-and-a-quarter-hour ride in Liège-Bastogne-Liège three days before, but there wasn't a hint of fatigue as they and another Gewiss teammate, Moreno Argentin, dominated Flèche Wallonne. By finishing first, second, and third, the Gewiss riders wrote a story too good to be believed. About fifty miles from the finish, they made a three-man escape from the pack and, though they were chased furiously, they were not recaptured. What astonished was the fact that such a small breakaway, just three riders sharing all of the pacemaking, could survive for so long. As members of the same team, they had every other team in the race battling against them. Still, they could not be caught. Argentin won, Furlan was second, Berzin third, and the fourth-place finisher, Gianni Bugno, was one minute and twelve seconds behind.

How could Gewiss be so strong? And it wasn't just that they dominated in the Ardennes—Furlan had won the Milan–San Remo classic a month before. Neither was it solely them—the overall prowess of Italian teams was astounding. Armstrong was the only non-Italian in the first five finishers in Liège-Bastogne-Liège, and the first eight in Flèche Wallonne were all Italian. If one looked at professional cycling through the lens of history, the 1994 results made no sense. And for those most affected by the Italian ascendance, the poor and vanquished opposition, there was now a dreadful dilemma. If you believed they were doping, as most did, and you knew the drug was powerful and undetectable, what were you to do?

On the morning after Flèche Wallonne, the cycling world was offered a definite indication of Gewiss's cheating. Michele Ferrari, doctor to the team, spoke publicly for the first time about r-EPO. Ferrari had worked under Professor Conconi at the University of Ferrara and assisted Conconi in the mid-1980s in preparing Italian cyclist

Francesco Moser for an attempt on the world hour record, an individual race against the clock in which the racer tries to go farther in one hour than the world record mark. Considered Conconi's star pupil, Ferrari wanted to be his own man, and after his involvement with Moser, he struck out on his own and began working with professional cycling teams. As reports of his impressive earnings from cycling wound their way back to the University of Ferrara, the rector was unimpressed. "Ferrari," said Professor Conconi, in a reference to his colleague's liking for financial reward, "was like a son who went the wrong way."

Gewiss's one-two-three in Flèche Wallonne had generated as much skepticism as admiration, and the former was very much the reaction of Eugenio Capodacqua, the Italian journalist with the daily newspaper *La Repubblica,* and his colleague Marco Evangelisti from the sports daily *Corriere dello Sport.* They were in Huy for that 1994 Flèche Wallonne and soon after the race ended, they saw Ferrari not far from the finish line. Capodacqua had enough sources to know r-EPO was the sport's new cancer, and he suspected Gewiss's riders were using it. So he and Evangelisti moseyed over to Ferrari, who was eating a slice of *saucisson.*

"Doctor," said Capodacqua, trying to gently needle his compatriot, "there's a lot of fat in that." Ferrari replied it wasn't a problem because there was a lot of goodness in it as well. They then got to talking about performance-enhancing drugs, and according to Capodacqua and Evangelisti, Ferrari said anything that wasn't forbidden was allowed. He also claimed it was worse for the athlete's health to train at altitude than to use r-EPO. Capodacqua asked if r-EPO wasn't dangerous and Ferrari claimed everything was dangerous if used unwisely. Capodacqua and Evangelisti believed they were listening to a man who was guided by what was and what was not detectable—and if it was not detectable, then it was permissible. "For me, he had the mind-set of a doper," says Capodacqua.

Jean-Michel Rouet, a journalist for the French sports paper *L'Equipe,* was intrigued by the story of the Flèche Wallonne race. Three riders from one team breaking away and withstanding the efforts of the entire peloton? Rouet had reported on professional cycling for fifteen years but had never seen anything like that before. What's more, he didn't believe that it was possible. Yet it had happened. "I watched how they escaped from the others; they were just riding at the front, not really trying to break away but riding so fast, they created a gap between them and the pack. They were from another world." Rouet wanted someone to explain and he made arrangements to interview Emanuele Bombini, the Gewiss directeur sportif, the following morning. Bombini had already departed for Italy when Rouet got to the Gewiss hotel, and Rouet was just about to leave when he noticed three Italian journalists talking to Michele Ferrari. "My Italian isn't great," says Rouet, "so I asked my driver to interpret, and we had only sat down when the conversation turned to EPO. 'I don't give it myself,' Ferrari told us, 'but if others are doing it, well . . . why not?' I couldn't believe what I was hearing."

In the following day's *L'Equipe,* Rouet wrote about the interview with Ferrari, who would later say the French newspaper had accurately reported his comments.

Jean-Michel Rouet: Speaking of EPO, do your riders use it?

Dr. Ferrari: I don't prescribe this stuff. But one can buy EPO in Switzerland, for example, without a prescription, and if a rider does, that doesn't scandalize me. EPO doesn't fundamentally change the performance of a racer.

J-M R: But EPO is dangerous. Ten Dutch riders have died in the last few years.

Dr. F: EPO is not dangerous; it's the abuse that is. It's also dangerous to drink 10 litres of orange juice.

What meaning would the Motorola riders have taken from this? Armstrong was in top form for the Ardennes races—the short sharp climbs in that part of Belgium suited him—and if the second place at Liège was hard to take, what happened at Flèche Wallonne was close to unfathomable. At the very moment the three Gewiss riders went clear, Armstrong was right there, alert to the danger and trying his damnedest to go with the three breakaways. Though he was only twenty-two at the time, Armstrong was the world road champion, blessed with considerable natural strength, and it was a given that if he went after three breakaways, he would catch them. That day he tried and couldn't close the gap. Something wasn't right. How then would Armstrong and his Motorola teammates have reacted to Ferrari's contention that he would not be scandalized by riders using r-EPO?

Criticism of Ferrari's comments in Italian newspapers was followed by his immediate dismissal from Gewiss, but it was a sacking designed to reassure the public rather than an attempt to lessen the doctor's influence in cycling. He continued working with the Gewiss riders but now in a private capacity, and the new notoriety helped to bring new clients to his office in Ferrara. That Ferrari was free to continue in cycling was obvious from the comments of UCI president Hein Verbruggen in the Dutch town of Maastricht two days after Flèche Wallonne. Among the journalists who spoke with Verbruggen in Maastricht was *L'Equipe*'s Rouet. "Someone asked Mr. Verbruggen about Ferrari's comments on EPO and he said it was 'just rubbish journalism,' he didn't believe Dr. Ferrari said those things. Verbruggen then talked about the relatively low number of positive tests in cycling and used this to suggest the sport didn't have a problem. It was hypocrisy."

Because of r-EPO's significant impact on performance, it was a big debate for cycling, and those opposed to doping felt they were on the losing side. "After the Ardennes races and Ferrari's comment that EPO wasn't dangerous," said Andy Hampsten, "there was an article in the Italian sports paper *La Gazzetta dello Sport* in which Bruno Roussel, directeur sportif of the Festina team, called on Hein Verbruggen to do something about the Italian teams who were using r-EPO. Bruno suggested unless something was done, it was a choice between allowing his team to start using the drug or losing his sponsor." Though coming to the end of his career, Hampsten experienced the pressure to use r-EPO. "It was like if you are a serious professional, you had to do a medical program, as they called it. You would have to do r-EPO just to keep up. It was presented to me by different individuals that if I was a serious pro who wanted his rightful place in the peloton, I should be doing r-EPO because I was being made a laughingstock."

The difficulty for Verbruggen and every other sports and antidoping official was that they didn't have a means of detecting r-EPO and so couldn't stop riders from using it. But to admit that would have been to concede that results couldn't be trusted. Instead, Verbruggen adopted a different position. He said the Italians were training harder and that the complaints about them were sour grapes; r-EPO wasn't the problem, rather it was laziness of those teams that couldn't beat them. "I believe we are still suffering the repercussions from that attitude," said Hampsten.

The Colorado rider wasn't the only one on the Motorola team uneasy about attitudes toward doping in the sport. At the 1993 Tour de France, in the season before r-EPO abuse became widespread, his teammate Steve Bauer slumped on a sofa in the lobby of Hotel Xalet Ritz in the Pyrenean village of La Massana and said something about never wanting to ride the Tour again. Two and a half weeks into the Tour, Bauer could hardly have been surprised to find himself so wasted, but there was an edge to his mood that was different. "See that boy?" he said, pointing to a two-year-old boy who toddled across in front of him. "I'd bet every penny I've earned that if you measured the level of testosterone in his body against mine, he would have more than me. Two weeks racing the Tour depletes you. Our team doesn't use testosterone but there are teams that do."

Perhaps the most telling commentary on the sport's doping culture from a Motorola rider came from Scott McKinley in 1994. McKinley had ridden for two years in Europe, one year with 7-Eleven (1989) and a second with Motorola (1990), and toward the end of the 1994 season, he joined an Internet newsgroup (rec.bicycles.racing) discussion about doping and posted his overview on Europe's drug culture. He wrote:

As a current pro who spent 2 years on the international amateur scene, 2 years pro in Europe with 7-Eleven and Motorola, and 3 years pro in the States, I feel obligated to offer up the sum of my observations on drug use in cycling.

I have seen directors of small kermesse teams [in Europe] making jokes while sticking needles into their riders in the dressing rooms. I have seen riders inject themselves during races. In the small races during the summer in Belgium and Holland, this practice was not uncommon when I raced there. . . . I remember the first summer I raced the pro kermesse circuit, in 1989 with 7-Eleven. Nathan Dahlberg and I were doing 4 or 5 kermesses a week and suffering to finish each one. Then suddenly the weather got a little hotter and our results began to improve. Nathan said to me that the heat ruins the effect of amphetamines, which was why all the riders suddenly seemed human. When the weather cooled again, all those guys with the thousand-yard stare who you've never heard of and who can ride tirelessly at the front for their leaders for 80k at 50kph reappeared. Hmm, I thought . . .

This is not to say that real races are exempt from drug abuse. What it means is that in races in which there is NO testing, riders use a whole different kind of drug; amphetamines, generally, which makes you ride completely differently than other drugs. . . . The untestable drugs such as: EPO, steroids when timed correctly, human growth hormone, cortisone, etc, have a much more subtle effect. These are what Tour-caliber riders use, if they use at all. Generally the program is very carefully monitored by a doctor or hospital, and there are usually no mistakes. The deaths in the early days of EPO were almost certainly because of careless, unsupervised usage. . . .

Now, who uses them? What is the percentage of abuse? The problem is that no one really knows except the athletes and the doctors or *soigneurs*. Even among the athletes them-

> selves there is little discussion. Drug use tends to be a discreet arrangement between an individual and his doctor. So if we go by the iceberg-tip philosophy that only the mistakes are caught, we can safely assume that a lot of people in the sport are using drugs, based on the 8 or 10 positives every year. My conservative, educated guess is that among the elite of the sport, at least half use banned performance-enhancing drugs and most of them get away with it. I think Lance and Andy really may be the clean exceptions among the top-50 riders in Europe, and even that may be wishful thinking. Unfortunately, it is simply impossible to say whether a rider is "clean" or not, no matter what your opinion of that person's character or how vehemently he defends himself. Only he knows for sure.

Two days later McKinley posted another message, this time in response to points raised in Thomas H. Kunich's posting:

> Regarding the benefits of EPO et al, I'm sorry I can't name names but while training in Italy last season, I visited a popular Italian sports doctor for a routine AT [anaerobic threshold] test whose clientele included Adriano Baffi, Max Sciandri, and a host of Italians and Russians. He told me and the person I was there with that getting on his full "program" (EPO, etc) would show an improvement of up to 30%, depending upon the physiology of the athlete. He offered this program to us matter-of-factly, the way an American nutritionist would offer a vitamin program. Needless to say, I declined. . . .

Following Gewiss's form in the spring of 1994 and Ferrari's public statement that r-EPO was easily available in Switzerland, more teams and riders turned to the drug to give themselves a better chance of competing. Festina's directeur sportif, Bruno Roussel, and the team doctor, Eric Ryckaert, knew some of their riders were using r-EPO, and to make doping safer and more effective, they decided to organize and administer a team doping program. Festina became France's top team. ONCE was the strongest Spanish team and it, too, was accused of doping its riders. Two Danish journalists, Olav Scannig Andersen and Niels Christian Jung, followed the 1995 Vuelta a España and after searching the medical waste left by ONCE doctor Jose Aramendi at a hotel in Orense on September 8, they found twenty-eight used syringes and six empty ampoules among the waste. The ampoules were subsequently analyzed and found to contain traces of r-EPO.

As the pharmaceutical race gathered pace through 1994 and 1995, Motorola lost ground. Their doctor wasn't prepared to organize and oversee a doping program. In fact, Max Testa tried to argue r-EPO wasn't the all-powerful drug it was cracked up to be and instead talked up the risks to riders' health. Testa, of course, suspected r-EPO was a significant performance enhancer and knew enough about how the drug worked to understand that if used carefully, the risks were minimal. What was a doctor supposed to say? He knew many of the Motorola riders believed that without r-EPO it was impossible to get good results. That did not mean he would support a doping program. Perhaps it was because he didn't see himself as a guru in the world of professional sport but more as family doctor to a group of mostly American sportsmen far from home.

When he started with 7-Eleven in the mideighties, Testa helped some riders find accommodation near Como in northern Italy, close to where he himself lived. Bob Roll, Jeff Pierce, and Ron Kiefel found a place in Brunate, just outside of Como. Andy Hampsten found another place just three or four miles out of Como. Lance Armstrong, Frankie Andreu, George Hincapie, and Kevin Livingston also ended up in the Italian city and Testa considered them normal patients; that is, he was there for them when they were sick. In an r-EPO-fueled peloton, that wasn't enough.

"In one respect," said Testa, "I was like an ostrich during these years [the midnineties]; there were things I didn't want to see. But at the same time I had to know what was going on and not be afraid to speak about it. If I said I don't know anything about it, the Motorola riders would have completely shut me out; I would have been a nobody. Sometimes they tested you out, tried to see what you knew, and I kept up-to-date because I wanted to maintain contact with them."

He remembers one specific conversation about r-EPO. "One in particular, don't remember which race, it might even have been a training camp. We were in this room and I had made copies of a couple of studies that were published on the effects of increased hematocrit and increased hemoglobin. This was the time Gewiss, who were rumored to have 60 percent hematocrits, were dominating the classics and I was trying to say r-EPO wasn't as influential as everyone thought."

Testa wondered if some riders on the Motorola team might be doping, but he felt trapped on two counts. He couldn't mistake his suspicions for proof, and when a rider told him what he was doing, he had to respect the confidentiality of the doctor-patient relationship. There was also the moral dilemma presented by the rider who came looking for advice on how to use r-EPO—that is, advice on the quantity and the timing of the doses. As much as the doctor might disapprove of the act, he couldn't refuse to explain how to use the drugs safely. The team found itself between the devil and the deep blue sea: by doing the right thing—running a clean program—Team Motorola diminished its ability to do what it was set up to do—win bike races.

With riders such as Armstrong, Hampsten, Phil Anderson, Alvaro Mejia, Sean Yates, and Steve Bauer, Motorola should have been one of the world's strongest teams, but results were disappointing. Armstrong wore the rainbow jersey of world champion through 1994 and didn't win one race in Europe. Mejia had been fourth in his first Tour de France in 1993 but dropped to thirty-first in 1994—and he was the best-placed Motorola finisher. Hampsten's physiological tests showed he was as strong as ever, but he could no longer make his class count in the mountains. He won a couple of small races in 1993 and finished eighth in the Tour de France but didn't manage to win one race through the final three years of his career. As for Steve

Bauer, his horselike strength didn't count for much in the r-EPO years.

Through the early months of 1994, the Motorola riders were confused. They knew they were good enough and yet they couldn't compete. They knew of the existence of r-EPO without fully realizing how prevalent it had become. They knew it improved performance but not by how much. Their feeling that the game had changed utterly came from things they witnessed in races. "I worked with the same doctor, Max Testa, all through my career," said Hampsten. "All my physiological tests were done by him and I didn't deviate in those years. My VO_2 Max was just what it had been, my horsepower was pretty much what it was in every other year, but in the midnineties—'94, '95 through '96—it became cruel. I don't want this to sound like sour grapes and I don't know that everyone who was climbing faster than me was on r-EPO, but instead of dealing with the same ten or so characters on the climbs, I was struggling to stay in a group of fifty riders. It is very easy to say I was thirty-three or thirty-four years old, physiologically going down, but quite honestly, I believe there was a huge amount of the peloton taking r-EPO and probably other drugs as well."

Frankie Andreu's memories of those years are not much different. "Early in the season you would go to the Ruta del Sol, after that on to some other races in Spain, and you would have Mario Cipollini and the sprinters just flying up these hills. I couldn't get over the hills with the damn group and you had these guys, who normally would never have been able to climb with the lead group, way in front of me. And I was like, 'Ho, this is messed up, this is crazy.' Then you had people talking about r-EPO, about riders with hematocrits of 60 percent and people not being able to sleep at night or having to start exercising in the night because their blood flow was slowing right down. The rumor was that their pulse meters were set to trigger an alarm when their heart rates went below a certain point, and by getting out of bed and exercising they stopped themselves getting a heart attack.

"You would tell yourself it was all rumor, then you'd see it for yourself. I remember my first Vuelta a España and we're on these killer climbs and the entire ONCE team is at the front, all nine guys, driving it, splitting the group up, and when it was down to just fifty riders, their nine were all still there. For me it was just, 'Whatever.' You had riders who shouldn't have been in the front group leading up the climbs and riders who should have been there who weren't. It was probably a division between r-EPO users and non–r-EPO users. It seemed like more and more were starting to join the program so that they could actually compete in races, instead of just hanging on."

In his book, *Prisonnier du Dopage* (Prisoner of Doping), the French rider Philippe Gaumont recalled the same debate taking place on the Castorama team in 1994 and 1995. Five weeks before the start of the '94 Tour de France, the Castorama riders spoke about drugs with their team doctor, Armand Megret. "It was May 27," wrote Gaumont, "we were at a race in Morbihan and the riders asked for a meeting with the medical staff. The Tour was approaching and there was a certain anxiety because the riders knew that without r-EPO, they would not be able to survive. I was a new pro, I hadn't yet started taking banned products but I was there, in my place, with all the riders as they spoke to Armand Megret. 'Now, the peloton has moved to a faster speed. The corticosteroids, they are all very fine, but they are not going to allow us to compete with other teams in the major tours. We must have r-EPO like the others. Can you give it to us?' " Megret refused and left the team at the end of the season. He was replaced by a doctor who would provide the drug, Patrick Nedelec.

In early 1994, a number of Motorola riders heard Italian teams were using a centrifuge to measure their hematocrits. This device was portable and had the capacity to analyze eight samples of blood at one time and quickly come up with hematocrit readings. The results gave the riders information about the thickness of their blood. If it was too thick, they ran the risk of dying in their sleep; too thin meant they needed to refuel with r-EPO. Their hematocrits should have been in the forty-two to forty-five range, but to compete at this time they needed to be in the fifty to fifty-five range. Knowing one of the Italians with the centrifuge, Motorola's Sean Yates—a very good rider who preferred to be an above-average *équipier* (team rider) rather than an average team leader—decided to have his hematocrit measured.

After taking a drop of blood from the top of Yates's finger, the Italian rider told him his hematocrit was forty-one. "You might as well go back to bed," he said. Yates told his teammates about his prospects for that day's race. By then the Motorola riders weren't expecting much anyway, but it was still sobering to have so simple and yet so scientific an assessment of how clean blood couldn't compete with r-EPO-injected blood. The incident also demonstrated the new sophistication being applied to the business of doping. Where once riders offered their arms or their backsides to a soigneur untrained in medicine, they were now self-medicating under doctor's orders and taking the necessary precautions to minimize the risks.

Jim Ochowicz, the team's directeur sportif, was generally mild-mannered in his management of the riders, but the man who convinces the sponsor to back the team cannot help getting emotionally involved. Ochowicz had, after all, staked his reputation on the team's ability to get results, and it was tough to see a good team struggle. Mostly, relations remained pretty good and the riders continued to call him "Och," but his frustration would occasionally seep to the surface and on a few occasions the volcano erupted. He told them they weren't working hard enough, weren't professional enough, weren't hungry enough. But really what could Ochowicz do? If he acknowledged r-EPO was the difference, what then could he say?

Some of the riders found it hard to accept the criticism because they were training hard and he was pretending doping wasn't an issue. He praised Axel Merckx for his ride in the 1995 Milan–San Remo classic (Merckx finished twenty-first but had been with the lead group throughout, and as he was in his first season with the team, it was a good effort), and the riders liked Merckx so they were pleased Ochowicz recognized his performance. But some months later they heard Merckx was working with Dr. Ferrari and had been for some time. Hearing that, they became more convinced than ever that Och's assessment of why guys rode well or badly was pure bullshit.

Within the team, there were suspicions about one or two of the riders. In 1993 Max Sciandri was the first Motorola rider to own a thermos that he often kept in the refrigerator on the team bus. No one could say for sure what was kept in Sciandri's thermos but it was known r-EPO had to be kept at a low temperature. Frankie Andreu believes that up to 1995, Motorola was predominantly clean. "As a team we were pretty innocent. Och didn't want to know about doping, didn't want to stay in the room if the subject was being discussed. Max said we didn't need the shit other teams were using. But that was Max. If you told him the pain had moved from your knee down into your calf, he would say that was good, it proved it was leaving your body. We hadn't a clue about how much r-EPO we would have to use, how often we would have to take it, and how dangerous it might be. All we knew was that it was expensive to buy.

"The thing about Lance was he had to be successful, he didn't want a career that was average. At this time, the whole thing bugged him, it ate into him. He would bitch about it all the time, 'This is bullshit . . . these guys are flying . . . I can't believe he's doing this. . . . I should be killing these guys.' He did not say 'I am going to get on a program,' he never said anything like that, but, man, he was frustrated, and I am sure it was as a result of that he decided to begin working with Michele Ferrari."

Chapter 5

IF YOU CAN'T BEAT THEM . . .

It is a November evening in Auckland, New Zealand, 2003, summer in the Southern Hemisphere. Steve Swart sits on the terrace of a restaurant overlooking the harbor. Yachts bob gently in their berths; over there is the yacht that won the America's Cup. There may be prettier places in the world but, for him, this city will do. Anyway, it's home now. Swart lives with his wife, Jan, and their four children in Howick, a nice neighborhood in Auckland's eastern suburbs. He works as a property developer—buys a place, does it up, leaves his mark, sells it on. Sometimes, if the offer isn't what he expects, he will lease it. And though he doesn't make millions, it keeps him content. What he likes is the freedom to spend as much or as little time on a job as he chooses. Mostly, that means early starts and late finishes, but if he wants a day off, he takes it. He gave some of his best years to professional cycling, and though it had its moments, he left the sport with a sense of having been cheated. His question is this: "If you took drugs during your career, how can you know how good you were?" People can think what they will but he's sure that if nobody had taken drugs, his career would have been better. But that's secondary. He had the right to know how good he was. Doping took that away.

The last two years were spent with the American team, Motorola. But to understand the end, it is necessary to know what came before. It began in the mideighties when he was a young man from a rural background on New Zealand's North Island, a young man who didn't want to be another forgettable face in a small country. A diesel mechanic by training and a bike racer by inclination, he didn't want a life that ended with *What if?* So he trained like a zealot and when the chance came to race in Europe, he was on the first flight out. Italy, France, Switzerland—he spent time in them all, lived in dingy apartments, survived on less food than he needed, endured the loneliness, and came home when the money ran out. In March 1987, Swart got his first break when a newly formed English professional team, ANC, gave him a contract. The £500 a month wasn't great, the work was hard, and six months after the team started the wages dried up. ANC used a Scottish soigneur, Angus Fraser, who would turn up at races and give injections to the riders. "You didn't ask any questions," says Swart. "You accepted it, as it was part and parcel of the team. You put your absolute faith in these guys because you thought they knew what they were doing. Did I think it was doping? I really didn't know. Because we weren't testing positive, you thought it couldn't be that bad. Yet I remember after the Tour of Britain, I decided to clean up my act and to ride being just me. No more injections from Angus. I just declined."

After a disheartening season with ANC, Swart joined the small Belgian team, SEFB, in 1988, and from England he moved to Liège. A year wiser, he quickly sussed out how things worked with SEFB: the team didn't have the resources to employ a doctor or a specialist soigneur like Fraser. Instead, it was up to each rider to organize his own medical backup, although the team's regular soigneurs would do what they could to help. "I remember going into the soigneur's room one night," he says, "and all this medical stuff was there. Basically, the riders were picking out what they needed. Another evening at the Tour of Switzerland we were sitting in a room when a soigneur came in with medications and everyone just helped themselves. There were guys just loading up, using great big syringes to fire stuff into themselves. Syringes like you might use on a horse." One of his teammates at SEFB was a young Belgian rider, Johan Bruyneel, who would go on to have a successful career, especially the years he spent with the ONCE team in Spain, but it would be as directeur sportif of the U.S. Postal Service team that he would eventually make the biggest impact.

Swart rode well with SEFB, especially in the 1988 Tour de Suisse, but by the end of that season he was ready for something different, and in 1989 he went to California, started as an amateur with a small cycling club, and worked his way up. They were good years, the racing was predominantly clean, no big syringes, no dodgy soigneurs, and Swart moved through the ranks so well that by the end of his fifth year, he was wanted by the best U.S. team, Motorola. Saying yes meant returning to Europe, but it wasn't a tough call because the European circuit was cycling's best, and even if it had a dark side, there was no racing to compare with it. He had gotten to the top of the tree in the United States but knew that wasn't very high, and the view was limited. His meeting with Jim Ochowicz was straightforward; the Motorola team boss wasn't offering as much money as he was earning with the Coors Light team, but Swart didn't mind taking a pay cut. What he wanted was the opportunity to race at the highest level and find out how far he could go.

Five and a half years passed from the time he left Belgium to his return to Europe with Motorola; he left Liège at the end of 1988 and returned to the old continent in early 1994. He had left before the arrival of r-EPO, and when he returned, it was everywhere. "In 1987 and '88, it was one world. In 1994, it had completely changed. The increase in speed was unbelievable, especially on the climbs. In 1988 I could hang in with the top ten on the climbs. Six years later and I had improved as a rider, but I am not close, man, not even close. They are not using the same gears as before, no one was using the small chain ring anymore. It was like *wow,* the level had gone through the roof. I knew straightaway this was something I was going to have to face. Some of the older guys in Motorola were a bit demoralized. I mean, we could go to a race like Tirreno-Adriatico at the start of the season in Italy and spend the week riding to our limit to hang in. Just to finish. It was crazy."

Knowing the odds were stacked against them, Motorola tried to be clever. They targeted smaller races where they hoped the dopers would not be primed. Alas, the dopers were everywhere. Before the 1994 Tour de France, the team picked out the team time trial as the one they could win, and they prepared better than any of the European teams, going as far as having a special pre-Tour rehearsal over the route. They finished a close second, missing victory by just five seconds. And in early 1995, according to Swart, the team made the decision that many others in the sport were making: if you couldn't beat them, you had to join them. "My memory," said Swart, "is that there wasn't that much debate about us using r-EPO in 1994; it was mostly the following season. After the 1995 Milan–San Remo classic, I went back to Como to train with Lance, Frankie [Andreu], Kevin [Livingston], and George [Hincapie], who were all living there, as was our team doctor, Max Testa. I hung out there for a couple of days, staying at a hotel. It was a bad time for us. We weren't getting the results we needed. We couldn't get that king hit, that big win. It was tough on the directeur sportif; he had to answer to the sponsors. If you don't get results, you don't get funding; if you don't get funding, you don't have a team. And the media were hammering us. Jim Ochowicz didn't want to do a doping program, but as riders, we could feel the pressure to deliver results. No one forced us to dope, but in the end, you were either in or out: you couldn't survive in the sport without doing it."

While in Como with his Motorola teammates, Swart recalls a conversation during a training ride. "We were talking about how we were going to deal with this situation. The general feeling was we had to take more control, we had to do something. We were just talking amongst ourselves, deciding what had to be done. Lance was very much part of the discussion and his view was that we had to do something. At the time the mood was swinging more and more in favor of getting on a program, and we pretty much agreed that anyone in our Tour de France team needed to be on a program."

The discussion was couched in the euphemisms that are so much a part of all discussions about doping. They talked of doing more, of being competitive, of "getting on a program," but Swart says it was perfectly clear what was being decided. Many members of the team, he says, decided to use r-EPO for that summer's (1995) Tour de France, and it was left to each rider to organize his own supply of the drug. In Switzerland to ride the Tour de Suisse, Swart went to a pharmacy and bought r-EPO. Unlike some other countries, r-EPO was available in Switzerland without prescription. Swart paid over six hundred Swiss francs ($330) for a box of r-EPO ampoules and says he asked Max Testa to advise him on dosage. He started taking r-EPO after the prologue to the Tour de Suisse and immediately suffered a downswing in performance. "What I realized then is that you can't just start taking r-EPO when you are racing. As it starts to work inside your body, the effort takes a lot of energy from you, energy you need to race. I should have started the program well before the race, when I was resting. It was just inexperience, not knowing how to do this properly."

His wife, Jan, was upset when he told her what he was doing, but she was aware cycling was not a clean sport and accepted the inevitability of having to do what the others were doing. As soon as the Tour de Suisse hit the mountains, Swart's lack of strength leveled him; two days before the end, he quit the race. With the Tour de France coming up, he feared not being able to start but stuck with his r-EPO program—a set number of units every second day—and hoped he would be ready. He did improve and through the early, frenzied stages of the Tour de France, the effect of the drug was undeniable. Because those early stages were flat and riders' energy levels were still high, the pace was insane—yet Swart hardly felt a thing. It was like the race hadn't even begun. At the end of a tough stage, his recovery was quicker than he had ever known. Yet it was also true that the new fuel didn't significantly change his results—it wasn't like he started winning. When his supply ran out a week or so later, he didn't bother with any more, as he felt what he had already taken would sustain him to the end of the race, and in any case, he didn't want to spend another six hundred francs.

That the team had moved from being a generally nondoping team to one that doped was, for Swart, demonstrated by a scene on the night before the second Pyrenean stage of the 1995 Tour. This was five days before the end of the race. They had gathered in one room, possibly Max Testa's, when it was decided to check the hematocrits of each rider. By this time, the team had its own centrifuge, and when each of the samples was taken and the numbers checked, Swart didn't miss the significance of the fact that most of the riders were right up to fifty or over. His forty-seven hematocrit, inflated by r-EPO, was one of the lower numbers for the team. He would later testify Armstrong's was one of those around fifty. Almost two and a half weeks into the Tour de France, those hematocrit figures should have been in the low forties.

From where Swart was standing, the decision to use r-EPO merely put Motorola on the same start line as everyone else, because by mid-1995, all of the big teams were using it. "Everyone was walking around with their own thermos—you could hear the sound of ice cubes against vials of r-EPO—and there were always riders in the hotels looking for ice to keep their r-EPO at the right temperature."

With some r-EPO in his veins, Swart didn't ride much better in 1995 than he had the year before, and he would feel better about his effort in 1994. For a time afterward he thought about his r-EPO experience and wondered if his body hadn't reacted well to a manmade blood-booster or if it was simply the case that he wasn't a good enough pharmacist. Certainly when he was a boy, he never imagined he would have to know a lot about pharmacology to be a good cyclist. After the Tour de France he rode a one-day race in England and a stage race in Spain, and he was then supposed to ride the Championship of Zurich. But on the night before the race, Jim Ochowicz told him the team was letting him go. And though he was entitled to be upset at the lateness of the rejection, which left him little time to find a new team, it didn't bother him that much. Part of him was ready to pack it in, the part that believed his very ordinary wages didn't justify the shit he would have to take to continue earning them.

The night before the Zurich race, he went around the rooms to say goodbye to his teammates. The visit to Armstrong's room was going to be difficult because he knew the decision to off-load him would have been Armstrong's as much as Ochowicz's. Still, he wanted to shake the hands of Sean Yates and Frankie Andreu, whom he considered good guys. He did that and then went to Armstrong's room. The door was ajar. Stepping inside, Swart realized his teammate was in the shower and shouted that he had come to say goodbye. Opening the bathroom door a little, Armstrong stretched out a hand. "See you later," he said. "Yeah, see you later," replied Swart. That was it. And with that, the Kiwi was on his way. Before the season ended and the last Motorola check went into his bank account, he rode the Tour of Holland and finally the Commonwealth Bank race in Australia. He probably would have won in Australia but he crashed on the second to last day and lost his chance. It was never meant to end romantically.

On the last night in Australia, he talked with Kevin Livingston, Bobby Julich, and George Hincapie, who, at that time, were all young and ambitious. They wanted to know more about r-EPO and whether Swart, who had been a professional for nine years, thought it was something they would have to do. "Look," he said, "if you're gonna make it in this game, you're going to have to do that stuff, simple as that." It wasn't what he wanted to tell them, he didn't believe it was what they wanted to hear, but to have told them anything else would have been dishonest.

Sometimes in the middle of an Auckland winter, a highlight clip from the Tour de France will appear on television and he will stop what he is doing and allow himself to be transported back. It's a short trip that turns up some good memories, but he doesn't delude himself: it was a world poisoned by doping and he, like most professional cyclists, was a victim. What he wouldn't have given for the chance to be on a start line where every rider was naked, just that man and his bike, nothing else. In that make-believe world, he thinks his career would have been far better than it actually turned out to be. But there's no way of knowing and, ultimately, no lasting bitterness because things have moved on. He enjoys reshaping and refurbishing houses, turning mess to order, turning an honest dollar. And in the middle of the night during those Auckland winters when live coverage of the Tour de France is shown on television, an alarm is set in his house. Off it goes at three or four in the morning and at that ungodly hour he can hear the patter of footsteps as a television is switched on. Logan Swart, teenage son of Steve and Jan, loves riding his bike and follows television coverage of the Tour with the passion of youth.

Father and son have not talked much about Steve's time as a professional cyclist, but the boy understands enough about the sport to know what it takes to finish the Tour de France, which his dad did twice. Prominent in the family living room is a photograph from the 1995 Tour de France showing the six Motorola finishers lined up on the Champs-Elysées. Lance Armstrong is in the middle of the group with the five-year-old Logan Swart perched on the handlebars of his bike. Steve wonders how Logan will handle the story of his dad's doping for that '95 Tour and thinks he should tell his son before it is made public. But it will be difficult; Dad is naturally quiet, so, too, is the kid, and there are lots of things they don't talk about. Dad says he is not sure he can bring himself to have that conversation.

Lance Armstrong and Motorola performed better in 1995. As well as winning the eighteenth stage of the Tour de France, Armstrong won two other European races, the San Sebastian Classic toward the end of the season and a leg of the early-season Paris-Nice. He also won the U.S. Tour DuPont for the first time, his overall victory coming with three stage wins. But if '95 represented significant improvement on '94, still it wasn't a year that fully satisfied Armstrong. In the fall of '95, he spoke with Eddy Merckx, whose bicycle company was a sponsor of the Motorola team, and they agreed Merckx would speak with Dr. Michele Ferrari and set up an introductory meeting between the Italian doctor and Armstrong. Ferrari had a very full client list and wasn't looking for new riders, but as Merckx was asking and as Armstrong looked interesting, Ferrari invited him to Ferrara in northern Italy for some testing. So began one of the most infamous patient-doctor relationships in the history of sport, a relationship that was to remain secret for six years.

Few of the Motorola riders were aware that Armstrong was planning to hook up with Ferrari and would not know anything had changed until spring training camp the following season. Andy Hampsten had left the team a year earlier but says that if he had still been there, he would have tried to dissuade Armstrong from working with Ferrari. "With a friend or teammate, I would have argued to the end against him having anything to do with Dr. Ferrari. I don't know him [Ferrari] very well—I'm sure I chatted with him—but from his reputation in the peloton and certainly his public statements after Flèche Wallonne, and his attitude to r-EPO, how it needn't be illegal because it was not dangerous, I would have been absolutely and completely against working with him. I can't imagine in 1994 anyone on the Motorola team even considering having him affiliated with the team."

According to a description of their meeting in Daniel Coyle's book *Lance Armstrong's War,* Ferrari's first impression was of a "very big, very, very big" rider. One takes from this the doctor's disapproval of Armstrong's body shape, which foretells of subsequent weight loss and the enormous improvement in performance that would take place over the following five years. Yet there is conflicting evidence about the extent of the weight loss. In a study of Armstrong's extraordinary improvement published in the *Journal of Applied Physiology,* Professor Edward Coyle presents scientific data on the rider's physiological characteristics gathered over a seven-year period. For the purpose of the study, Coyle tested Armstrong on five separate occasions and recorded the results. Body weight was one of the criteria, but the only direct comparison that can be made is between the figures for November 1992 and November 1999, as both weights reflect what Professor Coyle called Armstrong's "preseason" condition. In November 1992, it was 78.9 kilograms (almost 174 pounds). Seven years later, four months after the first Tour de France victory, it was 79.7 kilograms (almost 176 pounds).

His Motorola teammates did not notice any loss of weight in their leader when he showed up for spring training at the beginning of the 1996 season. On the contrary, they saw a bigger and more muscular Armstrong. "God knows what happened during that winter," says Frankie Andreu, "but Lance came back the spring of '96 and he was frickin' huge. He looked like a linebacker. It was 'Holy shit, man, he is big.' Obviously we all noticed it and he knew we did. He said something about Ferrari not realizing the effect the weight room was going to have. There are people I know if they go to the gym and start lifting, their bodies produce huge bulk. Kevin Livingston was one of those, he just blows up. But with Lance it was more than just seeing him big. I mean, he was big, but he could now rip the cranks off the bike like never before. He was dropping everybody and anybody in training, whenever he wanted to. And at Paris-Nice that spring, he was so frickin' strong. He and Jaja [Laurent Jalabert] had this battle which ended with them finishing first and second. Lance was unbelievably strong, a totally different person."

Chapter 6

THE HOSPITAL ROOM— PART ONE

If the first three years of Lance Armstrong's professional career had been generally encouraging, the fourth, 1996, was his best to that point. After his excellent performance against Laurent Jalabert in Paris-Nice, he won the Flèche Wallonne classic in April and four days later finished second in the oldest and perhaps most beautiful of all one-day races, Liège-Bastogne-Liège. It was appropriate he should win the Flèche Wallonne under the guidance of his new trainer, Michele Ferrari, because it had been in this race two years before that Ferrari's men dominated and the doctor's comments introduced the wider public to the existence of r-EPO in cycling. That Armstrong had moved onto a new level was apparent from his spring performances in Europe and his domination a month later of America's most prestigious stage race, the Tour DuPont, where he won the overall classification and five individual stages. Those wins lifted him to fifth in the world rankings and his subsequent victory at the low-key Fresca International classic in Milwaukee was his eighth of the season.

The season had its disappointments, too, as Armstrong felt unwell during the first week of the Tour de France when the weather was terrible. On stage six to Aix-les-Bains, he abandoned. Back in the United States, he prepared for the Atlanta Olympics but didn't produce the medal-winning performances that many expected. To him, sixth place in the individual time trial and twelfth in the road race were nothing. His season ended with a series of races in Europe—second in the Tour of Holland, fourth in the Leeds classic, fourth in the Grand Prix Suisse, second in the Grand Prix Eddy Merckx time trial. His last race of the season was at Baden-Baden, Germany, on September 14.

Success on the bike masked a serious threat to his life. Armstrong experienced the first signs of testicular cancer several months before more dangerous symptoms forced him to act. During the winter of 1995–96, he became aware that his right testicle was slightly swollen but convinced himself it was nothing to worry about. Even at the Tour DuPont, where his domination was absolute, he didn't feel 100 percent. "I was too exhausted to celebrate on the bike. My eyes were bloodshot and my face was flushed," he later said. At the Tour de France, his decision to quit on the seventh day was explained by respiratory problems. "I just couldn't breathe." He rode the Leeds classic and the Grand Prix Suisse with what he would later call "an almost generalized cancer." On September 18, his twenty-fifth birthday, he experienced a violent headache but still refused to accept there was a serious problem. Two weeks later, he coughed up blood in the bathroom of his Austin home and, finally, recognized that he needed medical help.

Most testicular cancers are detected by palpation—that is, by the patients themselves or their doctor feeling the swollen area. There are multiple symptoms: the presence of a mass in the testicle, a swollen or enlarged testicle, feeling of weight in the scrotum, pain in the abdomen or groin, pain or discomfort in a testicle or in the scrotum, blood in the urine, increased size or heightened sensitivity of the nipples. The existence of testicular cancer can also be proven by a blood test that measures the levels of three substances. These biological markers are alpha-fetoprotein (AFP), lactate dehydrogenase (LDH), and the hormone human chorionic gonadotropin (beta-hCG), and in the case of testicular cancer the readings for these markers will be abnormally high. In a healthy patient, the level of beta-hCG is normally between one and two nanograms per milliliter of blood and almost always under five. At the time his cancer was detected, Armstrong's beta-hCG level was enormously high. The figures vary according to the source, and even Armstrong himself has offered us three different figures: 52,000 ng/ml; 92,380 ng/ml; and 109,000 ng/ml. The first figure was quoted by Armstrong in an interview with the French sports newspaper *L'Equipe* in November 1996; the other two appeared in his autobiography, *It's Not About the Bike.*

It doesn't matter which figure one takes because all three are excessively high and it raises one obvious question: how was it that his enhanced beta-hCG was not detected in any of the drug tests Armstrong did through 1996? Human chorionic gonadotropin is a hormone produced by the placenta in pregnant women and its presence in urine is the standard proof of pregnancy. Enhanced beta-hCG levels in men are seen in cases of testicular cancer, but in its synthetic form beta-hCG is used by men to stimulate the production of testosterone. At the end of a cycle of anabolic steroids, the body's natural ability to produce testosterone will often be impaired, and athletes may use beta-hCG to restart their own supply of testosterone. Hence, the synthetic hormone beta-hCG is a Class A banned substance, and if an athlete is proven to have abused it, he faces a two-year suspension. The potential benefit of testing for beta-hCG is that an elevated level may be caused by testicular cancer and, through a routine anti-doping test, the disease may be detected at a very early stage.

Given his victories in the Flèche Wallonne and the Tour DuPont, his stage victory and second-place overall at the Tour of Holland, his fourth places in two World Cup races, and his second place at the Grand Prix Eddy Merckx, it is certain Armstrong was tested many times through 1996. Medical opinion varies on the length of time the testicular cancer sufferer will show elevated beta-hCG levels, but it is accepted that in Armstrong's case his levels of beta-hCG would have been enhanced for some time before the eventual detection of his illness. Even if one chooses the lowest beta-hCG level given by Armstrong, 52,000 ng/ml, medical opinion is that the level would have been raised for some months. "Fifty-two thousand units of hCG is a lot," says Jean-Bernard Dubois, professor of oncology at the Faculty of Medicine in Montpellier, France. "There is a mathematical relationship between the level of the marker and the extent of the illness. This is important. On the other hand, it is difficult to date the beginning of the cancer because it's a cancer that grows very fast. A range? Between two years and three months for this cancer. Less than a month is just not possible."

Comparable to anabolic steroids, beta-hCG stimulates muscle growth, increases training capacity, stimulates levels of aggression, and pushes back the fatigue threshold. The abuse of beta-hCG was identified in 1983 and has been detectable through urine analysis since 1987. Its use was prohibited by the International Olympic Committee (IOC) and the International Cycling Union (UCI) in 1988. That year, a report produced by Professor Raymond Brooks revealed hCG was being used by top British sports people. Though it was then on the list of banned products, it was not systematically tested for because no official threshold was established for a positive result. It was also the case that beta-hCG was not looked for unless it was known the testosterone level was already elevated.

"At that time [the 1980s]," says Jacques de Ceaurriz, director of the French national laboratory for drug testing at Châtenay-Malabry, "there was not an established threshold and I don't know if beta-hCG was being detected. What I do know is that at this time [1996], testing for beta-hCG was operational and systematic. But there is a difference between the intention to look for a banned substance and the reality. For an analysis to be valid, it must go through two stages: first, the banned product must be measured and analyzed in the laboratory; second, the conclusions must be reported and acted upon.

Perhaps the second point was not carried out. Moreover, the relevant body knew full well that his [Armstrong's] beta-hCG levels were not brought to light. The case caused profound embarrassment."

Implicit in De Ceaurriz's observation is the possibility that Armstrong's elevated beta-hCG level was picked up in the test but perhaps not reported by the laboratory, or if it was reported, maybe it was not acted upon by UCI. De Ceaurriz knows he understates the case when he speaks of the embarrassment caused. Inside the medical commission of the UCI, there was ample reason for panic. What would happen if Lance Armstrong employed attorneys to investigate the case, figured out why his elevated levels of beta-hCG had not shown up in drug tests, and then sought compensation from the governing body? This was a failure of doping control that could have cost him his life. Armstrong, however, was calm about the failure of the drug tests to detect his cancer in its early stages. In response to questions on this subject, he explained in a January 1997 interview with the French daily newspaper *Le Monde* that he knew beta-hCG was looked for in doping controls.

"I would like to know what my level was at the time of the control," he said in a reference to his drug test after the Grand Prix Suisse in August, six weeks before his cancer was detected. "If it's true the UCI keeps all the results, it should be possible to know where my cancer was at that time." Asked by the American journalist James Startt about the failure of the doping controls to identify his elevated beta-hCG level, Armstrong regretted this had happened. But there was no public explanation, either from the rider or the authorities, as to why the system had failed.

In 1996 the French laboratory at Châtenay-Malabry was under the direction of Jean-Pierre Lafarge. Interviewed by *Le Monde* on November 24, 1996, Lafarge was categorical on the question of beta-hCG. "Testing for this substance was systematic. The cases were rare, probably lower than one case in ten thousand. In Lance Armstrong's case, it is surprising that no trace of the illness was detected during the controls." The analysis of the drug controls at the Grand Prix Suisse in August 1996 was carried out by an IOC-approved laboratory at Cologne. Its director, Wilhelm Schanzer, told *Le Monde* that his "laboratory had the capacity to find hCG." At the time, the Cologne laboratory detected a slight abnormality in Armstrong's testosterone analysis but did not think the deviation suspicious enough to take it further. Results sent by the laboratory to the UCI showed Armstrong's test was negative.

The only official reaction from the UCI to the apparent contradiction of hCG not showing up in drug tests came from Anne-Laure Masson in an interview with *Le Monde.* Masson was then medical coordinator of the world governing body. "I'm perplexed because if the level of hCG was also high, Lance Armstrong should have tested positive, in principle. For now, it is inexplicable."

In May 1999, less than three years after Lance Armstrong was diagnosed with testicular cancer, the English footballer Alan Stubbs played for Glasgow Celtic against their great rivals, Glasgow Rangers, in the Scottish Cup Final. It was a disappointing afternoon for Stubbs, as his team lost 1–0. But things got a lot worse when he was later informed that his postgame urine sample contained the banned substance beta-hCG. This meant Stubbs either had testicular cancer or he had been using a banned performance-enhancing drug. The good news was that Stubbs had not cheated; the bad news was that he had testicular cancer. However, the early detection of the disease made the treatment more effective, and after undergoing chemotherapy, Stubbs made a complete recovery. Eight years on, he is still playing at the highest level of the professional game in England and has enjoyed a fine career.

Another athlete to have his testicular cancer detected after a drug test was the Australian field hockey player Greg Corbitt. A member of Australia's silver-medal-winning team at the Barcelona Olympics in 1992, Corbitt tested positive for beta-hCG three years later. He insisted he had not used performance-enhancing drugs, a claim vindicated by the presence of testicular cancer. In Corbitt's case, his physicians believed early detection of the disease may have saved his life.

Jonathan Vaughters rode with the U.S. Postal Service team in

1998 and 1999 and remembers speaking with Armstrong in January 1998 about the failure of the drug tests to show up his testicular cancer. "We were in the car. I think there may have been four of us—Lance, [U.S. Postal Service rider] Christian Vandevelde, [U.S. Postal Service masseuse] Emma O'Reilly, and me. Lance said he had spoken to Hein Verbruggen or someone at the UCI about it. The UCI knew this should have shown up in the tests and they knew Lance was well aware of this—the system had failed Lance and it could have been fatal."

While Armstrong's cancer was the event that overwhelmed everything else through the fall of '96, it was a season of much change. After many years of hanging on to his sponsor and his team, Jim Ochowicz was finally forced to admit defeat. Motorola exited the sport and while Ochowicz searched hard for a replacement sponsor, he didn't get one. Most of the riders were keen to stay with him and some hung on too long, lessening their chance of getting a good deal with a new team. Armstrong talked with François Migraine, head of the French credit company Cofidis, and they agreed to a two-year deal worth a little more than two million dollars to the rider. Three other ex-Motorola riders—Kevin Livingston, Bobby Julich, and Frankie Andreu—also signed with Cofidis. Switching to a French team also meant moving to France, and all four packed their things in Como and headed for Nice on the Mediterranean coast. Livingston and Andreu shared a two-bedroom place near the airport, Armstrong got a place on a hill not far from the beach, Julich also set up in the Nice area, and Tyler Hamilton, a young rider with the fledging U.S. Postal Service team, joined his compatriots in the Nice area. But thoughts of good times in Provence were put on hold as Armstrong fought life-threatening cancer. Livingston visited his friend at Indiana University Hospital and so did Andreu. It was a visit Andreu would not be able to forget.

The visit took place on Sunday, October 27, 1996. That morning Frankie Andreu and his then fiancée, Betsy Kramar, traveled from

Michigan to Indiana and checked into University Place Hotel, which is connected to Indiana University Hospital. Frankie and his fiancée got there late Sunday afternoon and departed on Tuesday morning. At this point, Armstrong was three and a half weeks into his battle against cancer. He did not look well, but that was understandable because three days before, he had undergone surgery to remove two lesions from his brain. That Sunday evening, a consultation between Armstrong and two doctors is alleged to have taken place in a room at the hospital—a question-and-answer dialogue between doctors and patient that would be fiercely disputed many years later.

Frankie Andreu cannot precisely recall when the consultation took place, whether it was Sunday afternoon or evening, but Betsy remembers it happening in the early evening. They had spent the afternoon with Armstrong, when a larger group of visitors had been present. They ate in the hospital cafeteria and Kramar spoke with Armstrong about her fiancé's habit of disagreeing with people for the sake of it. "If you say it's black, Frankie will say it's white, know what I mean?" she said. "Boy, do I ever," said Armstrong, and they laughed. Armstrong was bald, unnaturally thin, and frail. At moments when Frankie Andreu was talking with somebody else in the room, Kramar showed Armstrong gowns from bridal magazines and asked his opinion. She and Andreu had been engaged for six weeks and their wedding was just two months away.

They also went through the couple of hundred or so e-mails and cards that had been sent to Armstrong by well-wishers. Some were read aloud and many of the fans said they were praying for Armstrong, which Kramar found ironic, given the patient was unashamedly atheist. Perhaps it was because they wanted to watch the Dallas Cowboys' football game on television, or maybe it was because they just needed more space, but at some point they moved from Armstrong's bedroom to a bigger conference room. Kramar recalls leaving the bedroom, turning a few corners, and walking down a corridor and into a conference room. Lacking strength, Armstrong walked gingerly; an intravenous drip, set on wheels, was still attached to his arm.

The television was on in the conference room, showing the Cowboys against the Miami Dolphins. There was a bathroom immediately to the left of the door, some chairs, a table, a couch; some of the visitors chatted, others watched the game. In the end, they were down to six, other than Armstrong—Frankie Andreu, Betsy Kramar, Armstrong's friend and trainer Chris Carmichael, Carmichael's then girlfriend Paige, Armstrong's girlfriend Lisa Shiels, and Stephanie McIlvain, his friend from the sportswear company Oakley. It was the first time McIlvain and Kramar had met and they hit it off well. McIlvain's loyalty to Armstrong was unmistakable and reflected in her constant visits to Indiana through the first weeks of his illness. Now she had brought Armstrong a juicer and she talked with Kramar about how it worked. Lisa Shiels did not say a lot. A young engineering student with a seriously ill boyfriend and upcoming examinations, she read a textbook while the others talked and watched football.

According to three of the people present—Andreu, Kramar, and McIlvain—two doctors then entered the room, possibly a senior doctor with a younger intern. Kramar suggested she and Andreu leave but Armstrong said it wasn't necessary. Kramar would later say one of the doctors matter-of-factly asked Armstrong if he had used performance-enhancing drugs in the past, and in a quiet, almost weary, tone he said he had, and mentioned by name anabolic steroids, human growth hormone, r-EPO, cortisone, and testosterone.

Kramar says that as soon as she heard Armstrong make this admission, she thought of the man she was about to marry and was momentarily panicked by the possibility that he, too, was using these drugs. She indicated to Andreu that they should leave immediately as she needed to speak with him in private. Andreu motioned to her to calm down but quickly realized this wasn't going to happen. They left the room and no sooner had the door closed behind them than the interrogation began. Kramar wanted to know if he was doing the drugs Armstrong had admitted to doing. "I'm not marrying you if you're doing that shit," she said. He swore he wasn't. They went back to their hotel and the conversation continued: he insisted he wasn't doing the same things as Armstrong; she told him if there was any doubt in her mind she wasn't going through with the wedding. After all, Kramar believed Armstrong's cancer had to be related to the stuff he had been using.

Frankie Andreu and Betsy Kramar left Indianapolis on Tuesday, October 29, and in the days after arriving home in Dearborn, Michigan, Kramar talked with her mother, Betty, and her friends Dawn Polay, Piero Boccarossa, Lory Testasecca, and Vivian Hackman about what she had heard at the Indiana University Hospital. The soul-searching with Frankie went on for three or four weeks until finally things settled down. On December 31, 1996, she and Frankie Andreu were married.

Chapter 7

POSTAL GOES EUROPEAN

Emma O'Reilly left her native Dublin in early 1994, decided to pack her bags and take her chance in the USA. The Irish, especially the adventurous and the strong-willed among them, have always migrated, and O'Reilly had a good feeling about America. Dublin was fine but she needed space. She had trained as an electrician and, after that, qualified as a massage therapist. She'd also been a racing cyclist in her teens—not as good as her brother Norbert, but she was okay. Eventually she was asked to work as a therapist with the national cycling team, and for whatever reason she preferred that to actually competing. When she dreamed of America, she saw herself working as a masseuse with a team, traveling from state to state, seeing a vast country, and never getting bored.

Young, energetic, intelligent, and, in the end, lucky—because the green card that allowed her to work in the United States came one morning in the post. She moved to Boulder, Colorado, got a job with the Shaklee cycling team, and spent most of her first two years working with them. It was a low-budget operation with little potential for growth and O'Reilly wanted to be part of something bigger. In December 1995 she was interviewed for a job with the Montgomery-Bell team, which she knew was owned by a company called Montgomery Securities. The word on the grapevine was that the Montgomery team was destined to get bigger, a rumor given credence by the company's decision to fly O'Reilly from Boulder to their headquarters in San Francisco for the interview.

Though she is smart, O'Reilly didn't know everything. She got it into her head that Montgomery Securities was a burglar alarm company, and she intended to wear her best denim jeans to the interview. After all, they were looking for a masseuse, not a company secretary. A mechanic friend who was also called out to San Francisco for an interview told her Montgomery Securities had nothing to do with alarm systems. Denim jeans, he advised, would not be appropriate. So she found something respectable and dressed to the nines, landed in San Francisco, took a shuttle from the airport to the Pyramid building downtown, and after getting past security at the entrance, took one look around and thought, "Oh, Mother of God, this is no burglar alarm place." O'Reilly had walked into the offices of one of California's bigger financial players.

She was interviewed by the former U.S. cyclist and gold medalist from the 1984 Olympics, Mark Gorski, and a young man called Dan Osipow. Gorski was the new team's general manager and Osipow the team's media officer. It was clear they already believed in the team and, like clued-in salesmen, they persuaded O'Reilly that the U.S. Postal Service team was going to travel far. Gorski talked also of the team's founder and big boss, Thom Weisel, who was CEO of Montgomery Securities. Weisel's dream, said Gorski, was to build a team that would compete in the Tour de France. That was music to O'Reilly's ears because she instantly liked the thought of returning to Europe as a masseuse with a team in the Tour de France. The interview went well and she left the Pyramid building with two thoughts in her mind: she wanted the job, and no one was taking this guy Weisel lightly.

What O'Reilly didn't know was that Weisel's desire to have a successful cycling team stretched back almost a decade. It developed out of his own career as a competitor in masters' cycling and his relationship with the Polish immigrant Eddie Borysewicz, who had coached him. Eddie B, as he was called, was head coach to the U.S. cycling team that won an impressive four gold medals at the 1984 Los Angeles Olympics—an achievement tarnished by the later discovery that eight members of the team had, with the coach's support and approval, used blood transfusions to boost their performances. Though blood doping was not illegal at the time, it was unethical and, as practiced by the U.S. team, dangerous. "Because it was too late to remove and freeze their own blood," wrote the Olympic historian David Wallechinsky, "the eight U.S. team members went to a nearby motel and injected other people's blood, some of them relatives, some of them not. The two U.S. points racers, Mark Whitehead and Danny Van Haute, both became ill. Match sprinter Nelson Vails seemed to think the procedure was required of team members and was standing in line waiting for his transfusion when the team doctor, Thomas Dickson, informed him that it was optional and discouraged him from going ahead with it. A relieved Vails left immediately."

Five years after the L.A. Games, Weisel financed a team run by Borysewicz and called it Montgomery-Avenir. A year later the team attracted new sponsorship and changed its name to Subaru-Montgomery. Borysewicz was given a budget of around $600,000 and recruited better-quality riders. One of those hired was the raw but immensely strong teenager from Texas, Lance Armstrong. When the team went professional in 1991, Armstrong chose to stay amateur and prepare for the 1992 Olympics in Barcelona. Hopeful of being invited to compete in the 1993 Tour de France, Subaru increased their backing of the team to two million dollars but were disappointed when Tour bosses invited them to combine with the French squad, Chazal, to form one team. Subaru-Montgomery wanted a proper place or nothing. What might have been the team's biggest moment became the greatest letdown and at the end of the season Subaru walked away. That was 1993; and in 1994, the Montgomery team went into hibernation.

Weisel's ambition, though, never wavered. Through the financial commitment of his own company and sponsorship from a cycling equipment company, the low-budget Montgomery-Bell team was put on the road in 1995. Midway through that season, Weisel got lucky. The U.S. Postal Service, one of the world's leading employers with a staff in excess of eight hundred thousand, committed to a three-year deal to sponsor his team. Budgetary increases were built in for the second and third years. At last, the Tour de France was a wholly realistic target for Weisel. Once the sponsorship was confirmed, high-quality riders were recruited. Andy Hampsten, experienced and very successful, was the marquee signing. Though in the autumn of his career, Hampsten's name still meant something on the European circuit and with him in their ranks the new U.S. Postal Service team could not be ignored.

And, further down the pecking order, Emma O'Reilly got a job as masseuse with the new team.

Among her new colleagues was Dr. Prentice Steffen. Though his regular job was at St. Mary's Medical Center emergency room in San Francisco, Steffen is also a sports doctor and a former amateur cyclist. For the three years prior to 1996, he worked with the various Weisel teams and knew about the plans for the U.S. Postal Service team. Much of Steffen's life as a cycling doctor had been spent on the U.S. scene but he'd been to Europe often enough to know how tough it would be over there. But, really, they had no idea. Not Steffen, not O'Reilly, not even the directeur sportif, Borysewicz. Steffen has a vivid memory of the 1996 Tour de Suisse because it was his first major European stage race and they went there with such enthusiasm, like a European basketball team traveling across the Atlantic to play in the NBA. Steffen knew many of the Postal riders personally and liked quite a few of them. Their names have stayed in his head—

Andy Hampsten, Marty Jemison, Tyler Hamilton, Darren Baker, Eddy Gragus, Mike Engleman, two Poles named Dariusz Baranowski and Tomasz Brozyna, and a German named Sven Teutenberg. Nine starters but just three finishers—Baker, Hampsten, and Jemison—and the best-placed Postal rider was over an hour down on the winner, Peter Luttenburger of Austria. Steffen remembers it as the slaughter of the innocents.

It was hard for Steffen to believe so much had changed since he had been with the 1993 Subaru-Montgomery team in Europe three years before, but the increased speed of the peloton alarmed him. The difference was felt most when they tackled big mountain passes. Steffen liked the joke of the time that all you had to do when the pros were going uphill was turn your head to one side and you could kid yourself they were riding on a flat road; the speed was the same. So brutal was the Tour de Suisse that even the team doctor was pushed into service. "For one tough mountain stage, I was in the second team car, covering our guys who were dropped. Tyler Hamilton was off the back, riding just to make sure he was inside the time limit. My job was to jump out of the car and push him as far as I could. It was against the rules but Tyler had nothing left. After pushing him for a bit I would stop, jump back in the car, drive ahead, and I'd be ready to push him again as he came by. I kept that up for as long as I could, and man, I got a serious workout that day."

Poor preparation was part of the problem for the Postal team, as they were unused to European racing and they had come straight from riding the U.S. championships. But everyone knew it wasn't just that. Equally, if not more, relevant was the medical preparation. Most of the European teams were using the blood-boosting drug r-EPO, and it was the principal reason for the murderous speeds in the mountains. The longer the race went, the more the Postal team suffered. Happy to escape the torture, Baranowski and Brozyna left before the end so they could return to Poland and compete in their national championships.

Even cycling's governing body, the UCI, was now admitting the sport had a serious problem with r-EPO. Two years before, its president, Hein Verbruggen, rubbished suggestions that the drug was affecting competition, instead praising successful Italian teams for being more dedicated than their rivals. Then, in February 1996, Professor Guy Brisson, director of the IOC-approved laboratory in Montréal, believed he had found a way of detecting r-EPO in blood tests, and the UCI quickly agreed to fund the project. As part of his research, Brisson proposed taking samples of blood from riders at Switzerland's Tour of Romandie in May. The riders refused to participate, but after being reassured by the authorities that this was a research project and the anonymity of each sample was guaranteed, they agreed to give blood samples at the Tour de Suisse a month later. The research did not in the end produce a test for r-EPO but it did highlight the extent of the sport's problem with the drug. The analysis of the samples showed the average hematocrit at 46.11 percent. Data collected by the Scientific Anti-Doping Committee (CSAD) of the Italian Olympic Committee showed the average hematocrit for a different sample of five hundred elite athletes at 42 percent. That difference translates as a 10.2 percent increase in an athlete's oxygen-carrying capacity.

Two or three days before the finish of the race, there was a short, inconclusive conversation between Dr. Steffen and two Postal riders that, according to the doctor, would lead to the end of his involvement with the team. "It happened just outside the doorway of my room. I have a pretty vivid mental picture. We were in a small village. We had maybe been walking along and talking when they approached me, Marty Jemison and Tyler Hamilton. Definitely those two guys; Marty did almost all of the talking. 'We need to talk about the medical program,' he said. 'Okay,' I replied. I had a feeling about what was coming. 'As a team, we are not going to be able to get where we want to go with what we're doing.' There was a pause right there, an invitation to me to respond. 'Well, I think right now I am doing everything I can.' As far as I can remember, he then said that 'more could be done.' I said, 'Yeah, I understand, but I am not going to be involved in that.' Tyler may not even have said a word. I tried to make the point that we were an American team and that we simply needed time to get used to European racing."

Steffen's resistance to doping was deep-rooted and personal. In junior high and high school, he had started experimenting with alcohol and marijuana, hoping they would take away his sense of not fitting in, and under their influence he did feel more comfortable in company and he believed he was more attractive to girls. But the newfound confidence came at the price of addiction, something he could control only up to a point: it did not stop him earning a place at Vanderbilt University's medical school in Nashville, but while at Vanderbilt he injected himself with the heroin substitute Demerol, as well as morphine and other opiates. Because he was an avid cyclist, he kidded himself that he was doing okay or smiled at the thought that he was the world's fittest morphine addict. Afterward he worked in Tucson and under the Arizona sun he rode his bike wearing long-sleeved jerseys so people couldn't see the needle marks. The addiction, though, couldn't be hidden forever and when his life began to spin out of control, he checked himself into the Sierra Tucson treatment center and learned that he wasn't a bad person, just very sick. With the help of Narcotics Anonymous, Alcoholics Anonymous, and their twelve-step programs, Prentice Steffen got his life back in order.

When Jemison and Hamilton came looking for him to do "more," they were asking the wrong man.

Marty Jemison was one of three U.S. Postal Service riders who finished that brutal Tour de Suisse in 1996, although he was one hour and forty-three minutes behind race winner Luttenberger. Jemison remembers the race in Switzerland as the toughest of his career. One particular scene is tattooed onto his consciousness: a breakaway group has escaped from the peloton, general manager Gorski is screaming from the U.S. Postal Service car, the riders are hammering the pedals desperately trying to close the gap, their legs aching, lungs burning, but the breakaways are not coming back. And what Jemison remembers is how Gorski just didn't understand how fast the others were going, how hard the Postal riders were trying to catch them. "All he saw was the bunch of money he spent on his new riders and not one of them in the break," he said.

Jemison's memory is not so clear on the conversation with Prentice Steffen. "I don't have a memory of that conversation taking place," he says. Reminded that Steffen recalls it clearly, he concedes there may have been something. "I can't remember that precise conversation. I have a memory of asking Prentice for more B_{12} injections, stuff like that. If it was for anything, it would have been for that. He was struggling to give us that kind of stuff as often as we wanted it. My position was, 'Come on, Prentice, we need this.' He wasn't comfortable giving us B_{12}s. Maybe Tyler and I would have wanted Steffen to find out what was going on in other teams, just to see what we needed to do to become more competitive. A year later Pedro Celaya became our team doctor and I adored working with Pedro. The riders' health was the most important thing for Pedro. He understood the sport made very big demands and that it was his job to help us stay healthy." Hamilton, on the other hand, has denied that any such conversation with Steffen took place.

Prentice Steffen believes the discussion he had with Jemison and Hamilton in Switzerland was a turning point in his relationship with the team. The good rapport he enjoyed up to that point was replaced by coolness. Even on the journey back from Europe to the United States, he could feel their reluctance to speak with him. "I stood my ground, stuck to what I believed in. There is a Latin phrase that every doctor in the world is familiar with: *primum non nocere*—first, do no harm. If you help people take stuff like human growth hormone and corticoids for nonmedical purposes, you are harming them. I felt the team was either going to accept my position or I had signed my own death certificate, that they would get rid of me. Deep down, I knew it was the latter. Afterward they used me for some small races in America, but that was pretty much it."

Back in San Francisco, he called the team's headquarters, mostly looking for Mark Gorski, but his calls weren't returned. Within the team, people knew Mark didn't always return calls, and so Steffen persisted—more calls, more messages on the answering machine, then faxes, and still nothing. He felt he was getting the cold shoulder. Toward the end of October 1996, Gorski eventually left a message on Steffen's answering machine. It was what he expected: the team had no further need for his services. Pedro Celaya, a Spanish doctor, was recruited to take Steffen's place, but the departing physician was not prepared to go quietly. On November 4, 1996, he wrote a letter to Mark Gorski, who he believed had made the decision to fire him. The letter began:

> Dear Mark,
>
> I've had a week now to consider the message you left for me concerning your decision to use Johnny's doctor instead of me next season. I'm afraid you've seriously misjudged me and that you will need to reconsider your decision.
>
> I feel that, by my efforts with the team since the '93 season, I've earned the opportunity to provide medical support for the team as we enter a new and exciting phase of our development. Certainly my training, qualifications, experience, knowledge, and dedication cannot be questioned. So why your decision?
>
> As is my habit, I discussed the situation with two of my close friends. The explanation soon became clear. What could a Spanish doctor, completely unknown to the organization, offer that I can't or won't? Doping is the fairly obvious answer.

In his letter, Steffen went on to offer Gorski a way out. He could reverse the decision to change doctors and tell the new directeur sportif, Weltz, that the team preferred to stay with the American doctor they knew. If Gorski decided against this, Steffen threatened to go public about what he believed was the reason for his exit. Toward the end of the letter he wrote: "As the only serious American cycling team now, we have a serious responsibility to both our sponsor and to American cycling. Keeping our team and its reputation clean is a major part of that responsibility."

The response from Gorski came by way of Keesal, Young & Logan, a law firm with offices in San Francisco. In their letter, they said they would sue Steffen if he dared to make public his allegation. It was also stated that if any public statement by Steffen caused financial damage to the team, he would be responsible for that. Steffen spoke with an attorney friend in San Diego who said it was clear the team was serious in its threat and suggested he drop it. He talked with other friends who gave him the same advice. Persuaded by his friends' arguments, Steffen let it go. But the feeling of having been wronged remained.

Thirteen months after that Tour de Suisse, the U.S. Postal Service team rode the 1997 Tour de France. Riders that had been beaten into submission in Switzerland were transformed a year later, and all nine U.S. Postal Service riders who started the '97 Tour finished it. "I watched some of the mountain stages on television," said Steffen. "The Postal riders were comfortable. From their point of view, it was vindication of the decision to change doctors. What they said to me was probably right, they were not going to get where they wanted to go with me as their doctor. Completely right."

Though backed by the U.S. Postal Service sponsorship, the team in 1996 was disorganized. Eddie Borysewicz was a good coach but he struggled with the logistics of running a professional cycling team. General Manager Gorski was keen but inexperienced. The first training camp was at Ramona in southern California, where the team ended up eating at the Sizzler restaurant; the food came marinated in grease. Though Emma O'Reilly was new to the team, she wasn't naïve. Of one thing, she was sure: before the team would find its feet in the unforgiving world of European cycling, there would be casualties. Eddie B wasn't going to last because he didn't have the organizational skills. "Bless him," O'Reilly said, "but he was chaos."

Neither did she see Prentice Steffen surviving, because he wasn't the kind of doctor you wanted in Europe. She heard one of the riders complaining that all he had was vitamin C. And Emma O'Reilly, who didn't know much about the medical side, knew a team needed more than vitamin C. Besides that, she felt Steffen was too nice to survive in professional cycling. His focus was on looking after people rather than getting results, and he had a sensitivity you didn't generally see in the sport. Driving the team car, O'Reilly would often travel at excessive speeds, and when she braked hard, she instinctively stretched her right hand protectively toward her front-seat passenger. Steffen told her it showed her maternal streak. "There's not much maternal about me, Prentice, but thanks anyway," O'Reilly said.

More than any of the other doctors she worked with, O'Reilly liked Steffen, but she certainly sensed that he wasn't valued because he wouldn't do the things riders wanted him to do. Pedro Celaya came onto the team and the performances improved. He supervised the medical preparation for the '97 and '98 Tours and in each of those years the team finished with its full quota of riders. A mechanic from the Dutch team, Rabobank, told O'Reilly that the Tour performances proved Celaya was a very good doctor. By hiring Johnny Weltz to replace Eddie B as team manager, the U.S. Postal Service chose to tap into Weltz's experience as a successful rider on the European circuit. He knew the races because he had competed in them, he knew many of the riders because he had raced against them, and he had an understanding of what it took to be successful. He had experience that Borysewicz couldn't match, but again, it was more than that.

Weltz had been part of the European scene for a long time; he knew the people and understood the culture. Pedro Celaya was similarly comfortable in this environment. Keeping a pro rider healthy meant far more than waiting for him to get sick, and if there were products that helped to maintain hormonal balance, it was expected the doctor would know how they could be safely used. Most of the U.S. Postal Service riders had a good opinion of Celaya. "I quite liked Pedro," says former Postal rider Jonathan Vaughters. "I believe the riders' long-term health was an issue for him. He would speak to you about the possibility of certain drugs causing problems further down the line, when a rider was in his forties or fifties. He didn't feel it was his duty to make judgments on what products riders wished to use but to make sure they didn't kill themselves in the process."

The culture at the U.S. Postal Service team was changing. As much as the arrival of Weltz and Celaya signposted a new direction, the hiring of Jose Arenas was testimony of the team's commitment to doing things the European way. Arenas was employed as a soigneur, a "carer," whose primary duties included massage, preparation of food, lugging of suitcases from cars to hotels and back, and distribution of food during races. "I remember him as a lazy and not very good soigneur," says Vaughters. "When he gave you a massage, he didn't get in very deep, and from a rider's point of view, that wasn't very good. He smoked a lot, which didn't help, either."

Jose Arenas worked closely with Pedro Celaya and to those around the team they seemed good friends. Arenas and O'Reilly were together a lot because they were both soigneurs, and though she thought he was a nice fellow, she had reservations about his suitability for the soigneur's role. He was no good as a masseur and his personal hygiene wasn't the best. She disliked having him around the riders' food and it killed her to see the haphazard way he would wrap the food for the musettes (lunch bags), which were handed to the riders during races. Whenever he worked in the kitchen alone, he left it in a mess; when he was at a race, the ice buckets came back covered in mold, and she kept telling him about the danger of riders picking up bugs from food that was not refrigerated at all times. She might as well have been talking to the wall; he agreed with what she said and carried on as before. In the end, she realized his real role was as sidekick to Celaya and for the skill he had in giving injections.

Amazed, she watched as he performed with the assuredness of an anesthetist. So quick and so precise, she reckoned he'd put most doctors to shame. One evening at a small race in France, the Circuit de la Sarthe, he was preparing syringes before treating the riders, when she spoke about how long he had been giving injections. After all, to be as good as he was must have taken time. “Ah, it’s my job, it’s very important,” he said, and she bristled at the hint of pomposity. “Come on, Jose,” she replied, “you’re not exactly saving lives here.” Still, she couldn’t take her eyes off him as he went about his work, putting the needle into the vial, drawing back the plunger, making sure there was no trapped air, and then straight into the rider’s backside. In and out before the rider realized he had been injected. She marveled at his skill.

As to what came out of the vials and went into the cyclist’s body, it hadn’t taken O’Reilly long to realize that there are questions in cycling you don’t ask.

Chapter 8

THE LEADER RETURNS

Thom Weisel is a successful businessman. First with the investment bank Montgomery Securities and more recently as CEO and executive chairman of Thomas Weisel Partners, Weisel is one of the bigger financial players on the West Coast of the United States. He is also an art collector and a trustee of the San Francisco Museum of Modern Art. And to ensure that he remains fit enough to enjoy his wealth and interests, Weisel rides his bike. In the fall of 1997 Thom Weisel struck one of the great bargains of his life. For his modest but improving U.S. Postal Service team, managed by the specially created marketing company Tailwind Sports, he signed Lance Armstrong. What made the deal so attractive to Weisel was the price he paid for his acquisition. The basic salary was just $200,000 per annum, a trifling amount for a rider ranked fifth in the world the previous year. Sure, Weisel agreed to a bonus system of $500 for every

UCI-ranking point Armstrong earned up to a total of 150 points, and $1,000 per point thereafter, but if the rider performed well enough to make those points, everyone gained.

Armstrong went cheaply because of many people's fear that after life-threatening cancer he would not regain his old form. His two-year contract with the French team Cofidis covered 1997 and '98 but ended after one year, during which Armstrong's illness prevented him from racing in the team's colors. For that year, 1997, Armstrong was contractually entitled to $1 million and was actually paid $676,630, or just over two thirds of what he was due. He felt badly treated, while Cofidis felt they behaved honorably in paying him as much as they did. The haggling led to hostility and a parting of the ways. Armstrong sought a new team but wasn't in any danger of being run over in the stampede for his services. Few teams believed he would return to top form, and their reticence was Weisel's opportunity. How could the businessman lose? Whatever Armstrong did, the $200,000 investment would be justified by the publicity surrounding his comeback. Plus the rider had no intention of failing. He was hopeful of winning races, accumulating UCI points, and reminding Weisel that "bargains" can sometimes end up costing you.

The signing of Armstrong breathed new life into the Postal team. Emma O'Reilly was at the 1997 world championships in San Sebastian when she heard the news from her fellow Postal soigneur, Freddy Viaene. A Belgian, Viaene could barely contain his excitement. He had heard about Armstrong's strong personality and felt he wouldn't stand for the bad organization that had been evident during the team's first two years. Armstrong, he reckoned, wouldn't enjoy working with the directeur sportif, Johnny Weltz, and that thought pleased Viaene, who didn't like Weltz. As much as O'Reilly might have wanted to seem cool about Armstrong's coming, she couldn't disguise her own sense of excitement. She, too, knew of his reputation and expected he would drive the team forward. And if Armstrong's presence led to greater success for the team, that also meant more bonus payments for the staff.

They call the weeklong race from Paris to Nice "the race to the sun" in recognition of the race's progression to the warmer south of France. And as the race takes place in early March, it needs to find the sun. The weather was bitingly cold as the 1998 race commenced. Not only that, but a strong wind blew, forcing the riders into single-file lines as each attempted to find shelter behind the man in front of him. The '98 Paris-Nice was important for the U.S. Postal Service team because it marked Armstrong's return to serious competition. His fellow riders knew they shouldn't expect much—a few low-key early-season races showed he was a long way off top form, and he had been out of the sport for more than a year. Besides that, cancer survivors were not exactly common in the peloton.

George Hincapie was Postal's designated leader as the riders set off on the first leg of the eight-day race to Nice. That meant Hincapie, not Armstrong, was in the privileged role, and if he punctured, the others would wait for him. And what should happen? Hincapie punctured, and at a point in the race when the wind was blowing and every rider was stretched. The Postal team waited for Hincapie to get a new wheel and then they were off, in pursuit of the pack. It was cold and wet and a crosswind meant they funneled into a tightly packed group on the right-hand side of the road. Every so often, one of them would look up, hoping to catch a glimpse of the last carriage in the train of team cars behind the peloton, but mile after mile there was nothing. This chase went on forever, brutal for every one of them, impossible for the man just back from a life-threatening cancer. Many of the others didn't notice him slow and move toward the left-hand side of the road; they were still riding desperately to get Hincapie back to the peloton. Back at the team hotel that evening, Armstrong told teammates they didn't understand, and he was right about that. How could they know what it was like to race in the wind and cold after four courses of chemotherapy and eighteen months' rehab? They were sympathetic and felt pretty certain that they had witnessed Armstrong's last race. They wished him well, for they were genuinely saddened that one of sport's more ambitious comebacks had failed.

Armstrong went home to Texas and licked his wounds, and when the cold of the Paris-Nice day left his bones, he started again. He began by relaxing and waiting for his enthusiasm to return. When it did, he went to Boone in North Carolina with a former teammate at Motorola, Bob Roll, and his coaching friend, Chris Carmichael. They brought their bikes, did a lot of riding, and Armstrong recovered so well he contacted his U.S. Postal Service teammates. "I'm coming back," he said. "I want to do the U.S. pro championships." They knew this wasn't just talk because they could hear the old enthusiasm in his voice. He returned to Europe in June and won the Tour of Luxembourg. Driving away from the race that evening, he telephoned Weisel. Unable to contact the team owner, he left a message saying he had just won the Tour of Luxembourg and was already due a bonus payment. "Thom," he said, "you'll be hearing from Bill [Stapleton, Armstrong's agent]."

With the return of good form came an increase in Armstrong's authority. As Viaene predicted, the new leader didn't hit it off with the team's directeur sportif, Weltz, and most saw that as the beginning of the end for the Dane. The team had improved under Weltz but the organization remained substandard and Weltz was not popular with the team's support staff. Some riders felt Weltz spoke in riddles and struggled with the fact that Armstrong's record as a rider was better than his. As for Armstrong, he quickly convinced his teammates that he was born to lead. "The thing about Lance was that he wanted absolute authority," says Jonathan Vaughters. "If he lost a race when everyone was working for him, it was crushing for him. He could not accept defeat. It wasn't like a lot of teams who have leaders that want the team's support but rarely win. And they certainly don't apologize to their teammates when they don't win, which was the first thing Lance would do. He would say sorry for not fulfilling his end of the bargain."

Frankie Andreu looked out for Armstrong during their four years at Motorola, but when they teamed up again on the Postal team, the younger one had been through cancer and was even surer of what he wanted from his career. "He'd become a good leader. Attention to detail, unbelievable. I mean, we used to have cookies in the hallways of our hotels so we could snack whenever we felt hungry. Lance took those away and had them replaced by fruit. He was that kind of guy. It was like, 'Okay, listen, here's what we're going to do,' and he had a way of involving people in the decision, even though the decision was his. If he turned up at a hotel he didn't think was good enough, he would insist that all of us were moved to a better hotel. When he chose to be, he could be very inclusive. He would thank the staff if they did something to help out, thank those riders who did their job, and he was very good at getting the best out of people. Good at playing the role of team leader."

Emma O'Reilly was raised in Tallaght, a working-class suburb to the south of Dublin, and coming from a blue-collar background, she empathized with Armstrong's determination. Once she began giving him massages, their friendship developed. "I genuinely liked him. He would shoot me for saying this, but there is something vulnerable about him. You know it's because of baggage he's carrying. His father left before he knew him and he had a bad time with his stepfather. Because of this, he's on this mission to demolish every rival and anyone else who gets in his way. The sense of a man on a mission was there before his cancer. Definitely, it was. But the cancer gave him the impetus to be even more focused on what he wanted to achieve. He realized he had been given a second chance. Revenge was also part of the motivation; those European teams who said no to him in 1997, who thought he would never make it back, they really got under his skin. The little black book inside his head is thick with the names of those who turned him down. And once you're in that book, you never get out. So there were two missions: one, to rise above the poor white background he came from; two, to get even with those cycling teams who turned him down, and the only way to do that was to get results."

Armstrong's arrival may have accelerated the rate of progress but the U.S. Postal Service team was improving in any case. Outclassed in 1996, the team regrouped in 1997 and was better again in 1998. The quality of the riders had gotten better and the organization improved to a degree under Weltz. Those on the inside began to believe the team could become very competitive. And if much of the confidence came from Armstrong's return to form, another part of it came from Pedro Celaya's appointment as team doctor. Celaya understood the need for extensive medical backup in a way that Prentice Steffen never did, and he knew how to help riders survive in a tough sport made brutal by the fact of so many rivals using performance-enhancing drugs.

The challenge facing clean teams was apparent from the cargo of drugs seized by French customs officers from the car of Festina soigneur Willy Voet three days before the start of the 1998 Tour de France. Festina was the world's number-one team and Voet had just crossed the French-Belgian border when he was stopped at six A.M. on July 8. Among the drugs and medical products found were 234 doses of r-EPO, 82 vials of human growth hormone, 60 capsules of epi-testosterone, and 248 vials of physiological serum. Imprisoned, Voet confessed that Festina systematically doped its riders. Its directeur sportif, Bruno Roussel, later confirmed this was the case. Roussel was one of those who had called for action against Italian teams using r-EPO in 1994, and since nothing was done, he and his team organized its own doping program. The discovery of Festina's secret led to the French police's immediately targeting professional cycling, and generally, they found performance-enhancing drugs wherever they looked. With so many banned drugs being used, the sport's drug testing was shown to be completely ineffective.

By the middle to late nineties, no team could claim to be ignorant about what was happening in cycling. Riders transferred between teams each year, and though one team might have responded differently to the doping question than another, the information about what was happening was passed from team to team by the transient riders. For example, when Motorola ended its sponsorship in 1996, four of its U.S. riders—Armstrong, Andreu, Bobby Julich, and Kevin Livingston—joined the new French team, Cofidis. As they also hired Philippe Gaumont and Laurent Desbiens, two French riders at that time serving six-month suspensions for using anabolic steroids, Cofidis's position on doping was, at best, ambivalent. The team also signed the Swiss rider Tony Rominger, who for most of his career had worked with the controversial Italian doctor Michele Ferrari. If the ex-Motorola riders entertained doubts about some of their Cofidis teammates, it wasn't long before they saw the proof.

The presentation of the new team took place at the Lille Grand Palais on January 9, 1997, and all four American riders attended. This was a surprise to the Cofidis team bosses because they hadn't expected Armstrong to make the trip from the United States. A month before he had still been undergoing chemotherapy. After the formal part of the presentation ended, the French riders chatted among themselves, as did the Americans and other non-French. According to Philippe Gaumont in his book *Prisonnier du Dopage* (Prisoner of Doping), the French riders then started drinking, and after a while a few of them returned to their hotel, the Novotel at Neuville-en-Ferrain, where they injected Pot Belge, a highly dangerous recreational drug made from a mixture of cocaine, amphetamines, heroin, corticosteroids, and analgesics. They then returned to the Grand Palais, bringing some Pot Belge and syringes for other teammates.

Gaumont then described how the American and other non-French riders were spooked by the sight of their new teammates' bulging eyes, and how they quickly finished their Cokes and orange juices to return to their hotel. As for the partying French riders, they got back to the hotel as the Americans were waking the next morning.

Recreational drug use was not confined to the Cofidis team; all the same, the team ended up with a recreational drug problem. Cofidis was a new team, but their French riders had come from other teams where this culture was already in place. Neither was this solely a French phenomenon. In Italy, Marco Pantani's short life included both recreational and performance-enhancing drugs; the same was true for another great climber, the Spaniard José Maria Jimenez. Still, it was a shock to the American riders on the Cofidis team to realize that many of their French teammates abused Stilnox, a trade name for the prescription sleeping pill zolpidem. By taking Stilnox and fighting the first thirty minutes when the desire for sleep is almost overwhelming, the recreational user can experience vivid visual effects, decreased anxiety, even a mild euphoria. The effects are heightened by taking alcohol and Stilnox at the same time.

Cofidis's riders would take more than one Stilnox tablet; Gaumont admitted he would use four or five in one evening and with alcohol. The effect was to make them feel drunk and ridiculously good-humored. Their nonparticipating teammates would hear them partying in the corridor of the team hotel in the early hours of the morning, and occasionally, part of the fun for the Stilnox users was to run from room to room naked. It was an eye-opening experience for American riders who had never seen recreational drug use in the sport. Then there was the case of the Cofidis rider who suffered from serious psychological difficulties. Teammates who knew him said he was trying to deal with an amphetamine habit, and at one race in Spain, this rider got so down, he locked himself in his room and didn't appear for two days. Eventually he disappeared in the middle of the night and no one ever saw him again.

Philippe Gaumont admitted that during his six-month ban, he continued to dope, and that after the ban ended, the thought of not taking performance-enhancing drugs just didn't occur to him. He also wrote about the team's preparation for the 1998 Tour de France. "I believe that Vezzani [the Italian doctor, Mauro Vezzani, contracted by the team to work with the riders] turned up in May 1998. We were in the middle of our preparation for the Tour de France and he was busy with all the riders who would be in the team for the Tour. He had, amongst others, the Americans Bobby Julich and Kevin Livingston and the French riders Christophe Rinero, Laurent Desbiens, and, for sure, me. From Italy, he sent us r-EPO and growth hormones in express parcels after packing them in ice to conserve them. They arrived at our homes with instructions on how to use."

Because Gaumont was a recreational and performance-enhancing drug user, it wasn't difficult to dispute his version of events. And many did. Both Julich and Livingston denied receiving any packages from Vezzani. The Scottish rider David Millar, accused of doping and recreational drug use by Gaumont, called his former teammate "a nutter" but subsequently admitted using r-EPO when arrested and interrogated by French police. At the time when he allegedly received r-EPO from Vezzani, Livingston was working with Michele Ferrari. After police raids on Ferrari's house near the town of Ferrara in northern Italy, computer files were seized containing names and blood test results of many of the riders collaborating with Ferrari. There are multiple entries for Livingston, and the one closest to the 1998 Tour de France was made on July 7, four days before the start of the race in Dublin. On that date, Livingston's hematocrit was recorded at fifty, meaning the percentage of red cells in his blood was far higher than it would be for most people. And far higher than it was for Livingston seven months before, when Ferrari measured his hematocrit at forty-one. Such a variation in hematocrit is not normal. A year before, cycling's governing body had imposed a rule prohibiting riders from competing in a race if their hematocrit was fifty or greater.

Such a high reading did not prove r-EPO use but was an indication that the blood had been manipulated.

The U.S. Postal Service team insisted it ran a clean program and that all of its riders played by the rules. Under the terms of its rider contracts, any rider who tested positive was immediately fired, and if anyone inside the team had doubts about the team's ethics, there was reassurance in the calmness of the team doctor, Pedro Celaya. He matter-of-factly insisted that the Postal team had nothing to hide. Emma O'Reilly remembers a race in Switzerland in 1997 when the team was subjected to a surprise early-morning drug test. "Freddy

Viaene, our head soigneur, was going off the deep end, running all over the place. I am talking to Pedro, wondering what's up, and Pedro is just standing there, exuding calmness. 'I'm going to kill Freddy. He is making it look like we have something to hide, when, Emma, we have nothing to hide.'

"I was thinking 'This guy [Freddy] is a maniac, the guy is a moron.' At the time I didn't know much and wouldn't have known what was going on. For some reason Freddy lost it that morning with Scott Mercier, an American rider who wasn't interested in doing any kind of drugs. Scott was a bright guy, just wanted his one-season fling with the team in Europe and then head back home. He and I would look at the whole scene, and the way the Europeans saw cycling as a matter of life or death, and we would laugh. It was like, 'What planet did they come from?' But that morning while Freddy was running around like a blue-arse fly, Pedro stayed totally calm."

That early-morning drug test in Switzerland took place thirteen months before Willy Voet's arrest and the scandal-ridden 1998 Tour de France. Such was the consignment of drugs in Voet's car and the widespread publicity that followed, that it was like the veneer of respectability had been peeled back from the sport, revealing the rottenness beneath. Three days after Voet's arrest, the Tour de France prologue took place in Ireland, and as one of the U.S. Postal Service's team cars made its way from the center of Dublin that evening, the passengers were discussing the implications for the sport. "Frankie [Andreu] was holding court," recalls O'Reilly. "He said 'I'm glad Voet got caught. I'm sick of it. It's ridiculous. Every year it is getting worse and worse. Once upon a time you could do this race on spaghetti and water, but it's become impossible now. I hope to God this cleans things up a bit because it's getting scary.' Frankie said what a lot of guys were thinking."

Once the teams returned from Ireland to France, many became nervous about the possibility of police raids and searches. The police did search team hotels, and five Spanish teams quit the Tour rather than continue in a situation where they felt under suspicion and where the police maintained an intrusive presence. In the peloton, the resentment was widespread and expressed in the riders' refusal to race one stage, and as for the Postal team's cool doctor, Pedro Celaya, he wasn't his normal self.

"Pedro was like a cat on a hot tin roof," says O'Reilly. "It was really funny. The whole thing made me laugh. I mean, I was just sitting there and watching them squirm. Pedro was terrified. He was trying to make out that because he was a doctor, he was in trouble by the mere fact of being a doctor. That made no sense to me. To me it seemed he was worried about getting the rap for the products we had. I couldn't buy the line about, 'Ah, the poor innocent man, he's the doctor.' He would say to me, 'We're a clean team, Emma,' and I was thinking, 'Yeah, Pedro, you got nine lads around in the Tour last year, and you did that on spaghetti and water.' Yeah, right!"

On July 18, one week after the start of the Tour, there was a fifty-three-kilometer individual time trial from Meyrignac-l'Eglise to Corrèze in central France. At that moment, the sense of a race under siege was at its height. The previous day the Festina team had been thrown off the race, the police raided the ONCE team bus, and in another search, the police cut up a wheel of Parmesan cheese inside Polti's team bus. And still the race went on, commanding less time and space in the media than the searches. For the U.S. Postal Service, that individual time trial to Corrèze was a good day because one of its riders, the American Tyler Hamilton, produced the performance of his fledgling career to finish second, beaten only by the formidable German Jan Ullrich. Not all within the Postal camp recall that Saturday evening in Corrèze solely for the merit of Hamilton's performance.

"A fellow staff member told me," says O'Reilly, "that something like twenty-five thousand dollars' worth of medical products were flushed down the toilet of a camper van that the team was using. I can remember the big field where the camper was parked, and when I heard we had to dump it, it surprised me that with so many police around, we were still carrying that stuff. But then there were still cars toing and froing to Mercatone Uno [team of eventual Tour winner Marco Pantani], and despite all the raids, people were still doing things."

Frankie Andreu said he heard something about products having to be flushed down the toilet but cannot remember from whom. Another Postal rider agreed to speak of this incident on condition he would not be named. "The story is true," he said. "I was in the camper when the stuff was flushed away. The gendarmes were all over the field, and at one moment it seemed as if they were all moving toward our camper, and so the stuff was flushed down the toilet. It definitely happened."

From their first meeting at training camp, Lance Armstrong and Emma O'Reilly hit it off well. Maybe it was the seriousness with which she went about her work that he admired, or her penchant for saying what was on her mind. He would have enjoyed her lack of affectation and the ease with which she moved in a predominantly male world. And she was a very good massage therapist, by far the best on the team. So as Armstrong's first season with the U.S. Postal Service began, no one was surprised the team leader wanted his massages from O'Reilly. When he recorded his first victory after the comeback from cancer, the Tour of Luxembourg, O'Reilly was his soigneur.

Though theirs was a good relationship that would get better, Armstrong had a generally low opinion of the Postal soigneurs, and when he made this point to O'Reilly, as he often did, part of her wanted to slap his face. The other part knew he was right. As a team within a team, the Postal soigneurs were average. But she also felt that part of the problem was that the soigneurs, especially O'Reilly, didn't want any involvement in the medical program. One time Armstrong was asking around for the telephone number of his old soigneur at Motorola, John Hendershot, and O'Reilly felt it was as if Hendershot were a real soigneur, not like the lot Armstrong now had to deal with. "You call yourselves soigneurs," he would say, and O'Reilly would tell herself to take a deep breath and remember that the criticism was not personal.

From her first year in professional cycling, O'Reilly had made the decision not to get involved in the medical programs of the teams she worked for. Traditionally in European cycling, soigneurs cared for the rider's medical needs. A multivitamin injection into the backside, a shot of cortisone, the right blend of corticoids, a glucose drip, a few vials of r-EPO, a saline drip, a shot of testosterone—the good soigneur could do all of these things and more. O'Reilly wanted nothing to do with it. She wasn't qualified to give injections and didn't want to involve herself in what she saw as doctors' work. Furthermore, she believed these "medical programs" to be morally questionable and wanted to leave the sport with a clear conscience. Yet, from her position on the inside, she could understand Armstrong's complaints.

"Part of me thought, 'I don't do the medical program, therefore I am not in the strictest sense a soigneur, not a proper European soigneur, not a soigneur like Willy Voet.' I didn't mind Lance complaining. At the Tour of Luxembourg in 1998, Lance was giving out about us; there was only myself and a Russian guy, Alex, who was chiropractor to another Postal rider, Viatcheslav Ekimov. Neither of us had a clue about the medical program. There was stuff in the truck but I didn't know what it was for. So I would open the truck and say, 'Here you go, lads, help yourselves.'

"I felt bad they had to look after these things themselves. I am sure complaints went back to America. This was part of my job description that I wasn't able to meet. What a useless soigneur. But it was too late for me to start learning, and anyway a voice inside my head kept saying, 'No, don't get involved, this is not your problem.' I had always known I was going to quit the sport sooner rather than later, and when I left I wanted to go with my head held high."

If Armstrong sometimes made her feel she was an inadequate soigneur, the flip side for O'Reilly was that it was he who convinced her that the team was going places. From the moment he turned up at the training camp at Ramona and began saying what needed to happen, the organization and the ambition moved up a level. O'Reilly was also impressed by Andreu. "I liked Frankie on the team; he was a very good addition. Frankie was a really good professional and a great road captain. A spade was a spade with Frankie; well, for the most part. If he had something to say to you, he said it and didn't mess about. He was easy to deal with because of that. I used to call him 'Ajax' because he was so damn abrasive. Think of the blue, powdery stones in Ajax, that's Frankie. He loved that name. 'She calls me Ajax.' He took it as a compliment. I meant it as a compliment."

By winning the Tour of Luxembourg, Armstrong showed his teammates he could be competitive again. It was a good race for the Postals, as the team rode well to protect their leader. "Helping Lance to win that race was a highlight of my career," said Marty Jemison. "I recall one particular stage where we were going into a town, and after the town there was a really steep two-kilometer climb. Tyler [Hamilton] and I rode at the front. Over the top, it was Tyler, Lance, and I. Lance wasn't superstrong; you could say he was a fragile leader, definitely not unbeatable. But the team was very good; his *domestiques* did the work for him. Two weeks later we went to Germany for another stage race and by now you could tell he was getting back his confidence and he rode stronger. And he won again."

After Armstrong won in Luxembourg, he, Andreu, and O'Reilly traveled together to Metz and pulled in at the Hotel Campanile on the outskirts of the city. Andreu and Armstrong waited in the car while O'Reilly checked if there were rooms. By the time she returned, the two riders had gotten into an argument with an irate Frenchman. The fight began after Andreu threw some orange peel out of the car window and upset the civic-minded Frenchman. He complained, and as it turned out, it wasn't the best moment for a civilized difference of opinion: the Postal riders had endured a tough final day in Luxembourg. Armstrong had won the race, Andreu had won the last leg—all they wanted was something to eat and somewhere to lay their heads. The argument soon became heated and the Frenchman was told where to go. "Can I not leave you for five minutes?" O'Reilly said when she came back.

The good news was that the hotel had rooms. They agreed to meet in the restaurant five minutes later. Over dinner Armstrong explained how he had rigged up his room with a trap so that if the French guy came back he was in for a surprise. As soon as he opened the bedroom door, he would bring Armstrong's homemade contraption down onto his head. O'Reilly and Andreu laughed so hard tears came to their eyes. "You're ridiculous," they said, "completely mad," but that was Armstrong, anticipating his enemy's next move, readying himself for the attack, making sure he wouldn't be caught off guard. That night, before going to sleep, he fixed a chair against the door so that if the Frenchman tried to force his way into the room, he would wake his intended victim. "A freak," said O'Reilly the next morning, "you're a paranoid freak."

O'Reilly's desire to stay clear of the team's medical program was respected by the team, even if there were those times when she felt Armstrong would have preferred to have her involved. Though she was on the outside, O'Reilly was aware of some of what was going on around her. She knew how much medical stuff was carried in the truck and how the team had black bags and white bags for carrying the stuff into their hotels. They bought thick white bags—plastic but not transparent, and they were used for the smaller packaged drugs. The bigger black bags carried bulkier products, syringes and IV equipment. When riders wanted products, they were put in a black bag and taken into the hotel.

O'Reilly noticed the secrecy and knew that not all of the products in the truck were legal. At different times she saw the actual boxes but recognized few of the drugs, as they were packaged with their brand names. She recalls the name Knoll Pharmaceutical Company because her uncle, a pharmacist, had worked for Knoll. She observed the comings and goings at the Postal team hotels, the stranger who had driven from Spain and turned up for no apparent reason. And there were other hints. At a small race, Celaya, Jose Arenas, and Freddy Viaene might all be absent, and as no one else on the staff wanted to give injections, the riders injected themselves. That was something O'Reilly came to regard as totally normal. She says Armstrong used to joke that he liked a room with pictures on the wall, so there would be hooks from which he could hang his IV bag.

And if Emma O'Reilly hadn't picked up the vibes along the way, then there were her own direct experiences with cycling's medical culture.

Chapter 9

THE PROGRAM

In having a system of medical support for their riders, the U.S. Postal Service was like all other elite professional cycling teams. Because the team was better funded than most, its pharmacy was better stocked and its riders subjected to more treatments than those on other teams. French law insists that anyone bringing medical products into the country must first receive authorization from the authorities. For the 2000 Tour de France, the Postal team was granted permission to bring 126 different medical products into the country. Twelve of the products, predominantly corticosteroids, were banned under anti-doping legislation and permissible only if a rider had a therapeutic exemption. According to official sources, Postal's medical arsenal was the largest declared by any team in the 2000 Tour de France—twice the amount declared by any French team and a third more than Italian teams.

After spending two years with the U.S. Postal Service team, Jonathan Vaughters was surprised by what he found at the French team, Crédit Agricole, in 2000. Crédit Agricole didn't carry anything like the amount of drugs and medications he had seen on the Postal team, and its riders received very few injections, which was not the case with the U.S. team. What was obvious to Vaughters was that the French team had responded positively to the scandal of the 1998 Tour de France and were trying to create a culture in which riders were treated by doctors only when they were ill or genuinely in need of medical care.

Emma O'Reilly saw Postal's stock of medical products—the syringes, the intravenous drips, the boxes of pills, the vials of liquids—and observed the sport's *omertà,* its unwritten law of silence. She didn't know what was legal and what was not, only that it was a subject not discussed with strangers. Even in conversations with fellow staff members, O'Reilly would refer to the "medical program" without using the word *doping.* After leaving the sport in 2001, she was asked what she meant by the team's "medical program." She wrote the following: "To me, it's the admission of drugs, both legal and illegal, for recovery and performance-enhancement."

There were times when she felt part of the program.

Served with a subpoena and forced to testify in a U.S. lawsuit involving Lance Armstrong at the beginning of 2006, she recalled the occasion in 1998 when one of the riders, George Hincapie, asked her to collect something from the Belgian soigneur Freddy Viaene. Hincapie knew O'Reilly was going to be at the Hotel Nazareth in Ghent and that it would be convenient for Viaene to drop it off there. Viaene stopped by the hotel, met O'Reilly, and handed over the package. "He told me it was testosterone because he didn't want me having it in my possession any longer than was necessary. He definitely didn't want me traveling anywhere with it. I passed it on to George the next time I met him, which was a day or two later." Testosterone, a hormone commonly used by cyclists and other athletes, is a Class A banned substance.

Late in the summer of '98, Armstrong rode the Tour of Holland, and after winning races in Luxembourg and Germany, the Postal leader continued to show good form by finishing fourth in the Dutch race. Coming late in the month of August, the Tour of Holland takes place at a time when teams are recovering from the previous month's Tour de France. No one who has contested the French tour has much enthusiasm for Holland, and the same is true for support staff. So the U.S. Postal Service's directeur sportif, Johnny Weltz, sent his deputy, Denis Gonzales, to Holland. Neither did the team doctor, Pedro Celaya, make the trip to the Netherlands. O'Reilly didn't mind being there because, with Armstrong riding well, the team was competitive. In terms of her working relationship with the team leader, it was important she was there because of Armstrong's complete disdain for Gonzales.

On the final day of the race, Gonzales was supposed to get Armstrong a lift to the airport but didn't. O'Reilly offered to take him and so they set off, just the two of them in one of the team's Volkswagen Passats. At the airport Armstrong handed O'Reilly a small, tightly wrapped plastic package. He told her it was stuff he didn't want to leave in the hotel room and would she mind dumping it somewhere along the route to her next destination. Inside the wrapping were empty syringes that Armstrong had used during the Tour of Holland. "Yeah, fine," she said. "I'll do that, no problem." It should have been a straightforward chore. . . .

Conscious the car was covered in U.S. Postal Service livery, O'Reilly thought about where she would dump the package. She didn't know what the syringes had been used for, and it may well have been that they were for the injection of legal products, but still, she wasn't prepared to pull in at the first service station on the motorway and put them into a trash bin. This was just four weeks after the end of the 1998 Tour de France, with all of its scandals and searches, and the risk of someone seeing her dumping the rubbish was too great. She thought the best thing would be to keep the package in the car until she arrived at Julien DeVriese's house in Ghent,

Belgium. DeVriese was Armstrong's mechanic on the Postal team and he and O'Reilly were good friends. Most times O'Reilly was traveling through Belgium she called to see Julien and his wife, Vera.

After crossing the border from Holland into Belgium, O'Reilly was cruising along at what for her was a sensible speed, perhaps a little over the official limit but deliberately restrained. After driving for some time she noticed the flashing lights of a police car in her rearview mirror and thought, "Oh shit." The officer wanted her to stop and all she could think of was the package. Seeing a slip road off the motorway just ahead, she pulled in there and he pulled up right behind her. By now O'Reilly was shaking, wondering how many syringes were in the package and what might be discovered if they were analyzed at a laboratory. Should she immediately admit she was carrying them or take the chance the car wouldn't be searched? In the back of her mind was the thought that the U.S. Postal Service livery was an invitation to the police officer to conduct a search. As the officer got out of his car and approached, she felt herself beginning to perspire. Her last thought was to play things cool and apologize for driving a little fast.

"Officer, I am very sorry for—"

"No, no, not that at all."

"Oh?"

"You work for the U.S. Postal team?"

"Yeah, that's right."

"Do you know Mark Gorski?"

"Yeah, he's my boss."

"I used to race with Mark in the eighties."

"Ah, is that so? That's interesting."

"Do you know how I could get in touch with Mark?"

"Yeah, no problem. I have Mark's number. Here, you want Mark's number?"

"That would be great. My son is now racing and I would like to speak with Mark and invite him to come and stay with us when he is here in Belgium."

"I'm sure he would like that. Look, you call him on that number. Next time I see him, I will tell him about meeting you."

"That would be good. Thank you."

"No, officer, thank you."

Before the conversation ended, O'Reilly and the cop talked like long-lost buddies and parted on great terms. The package was still safe and unopened in the glove compartment of the car. She didn't know what the syringes had been used for but neither did she want anybody finding out. When she arrived at DeVriese's house in Ghent, she told them the story of the eventful journey, and Julien, Vera, and their son Stefan almost cried with laughter.

Everyone could see that, deep down, Johnny Weltz wasn't a bad guy. In October 1996 he had been made directeur sportif of the U.S. Postal Service team, and for the following two years, that was the job he did. Weltz was from Denmark and had been a very good rider, and his appointment reflected Gorski's view that to beat the Europeans in their own backyard you had to properly know the sport. Weltz knew what it took to compete effectively in Europe. Many of the riders thought he was okay and could see he was clearly a riders' man. They respected his achievements and his understanding of the demands on them, even if some of the riders thought he was too soft. The team improved in 1997 and 1998, the two years Weltz was in charge. But there were difficulties. Weltz didn't get on well with the support staff—that is, the soigneurs and the mechanics. He made decisions they didn't agree with and they got nowhere when they suggested alternatives. While results did improve under Weltz, the team's organization remained substandard. These were wrinkles that might have been smoothed over if Weltz had been able to meet one imperative: get along with Lance Armstrong.

That was beyond Weltz, who just wasn't to Armstrong's taste. Whereas the Dane would come to things in a roundabout way, Armstrong got directly to the point and wasn't patient with those who didn't. If Weltz had been a brilliant organizer or a clever facilitator, he might have had a chance. Instead he was his own man, not a bad fellow but not the directeur sportif Armstrong wanted. There were achievements: the team had competed in its first Tour de France in 1997, fulfilling Thom Weisel's dream; and since the demise of Motorola, the U.S. Postal Service team had become the standard-bearer for American cycling. And Weltz might have argued that his less-than-warm relationship with Armstrong hadn't stopped the team leader riding strongly through his comeback year in 1998. Armstrong had won the Tour of Luxembourg and the Rhineland-Pfalz race in Germany and had also finished fourth in the Vuelta a España.

But if Lance didn't want Johnny, then Lance's results were not going to save Johnny. In the end, Weltz made it easy for Armstrong. By continuing to use his friend Denis Gonzales as assistant directeur sportif, Weltz damaged his own credibility. The riders and staff called the assistant director "Speedy" after the cartoon character because whatever Denis did, it wasn't in a hurry. Even the guys who liked Johnny couldn't understand from where he had gotten Speedy. By the middle of 1998, Armstrong knew the team had to ditch Weltz.

By the time they got to the Vuelta a España in September, he and Weltz were no longer speaking. There was no question about whose side the team would take, and Armstrong was already searching for a replacement. With his relationship with Weltz broken down, Armstrong confided more in O'Reilly during his evening massages. It was a good time for the Postal leader because Spain was the first "grand tour" in which he contended for overall victory, and his fourth-place performance made him excited about the future. During that time he and O'Reilly were getting on really well; he was happy to talk to her about his plans and ready to listen when she spoke.

"During the Tour of Spain we spoke about my situation. I wasn't happy with my salary—thirty thousand dollars—and I was looking for an increase. When I told Lance of my intention to leave if the team didn't cough up more money, he said, 'Fine, Emma, but make sure you come back to me before you do.' "

In Spain, Armstrong started to speak of Johan Bruyneel. The Belgian was in his final season as a rider with the Spanish-based ONCE team and he impressed Armstrong when they spoke. They agreed to meet up, and one evening Bruyneel turned up at the Postal team's hotel for a meeting with Armstrong. If it wasn't a formal job interview, it wasn't far off that. After the meeting, Bruyneel wrote an e-mail that excited Armstrong enough for him to tell O'Reilly he thought Bruyneel was "pure class." In the e-mail, Bruyneel said he could see Armstrong in the yellow jersey on the Tour de France podium in 1999, and also in the rainbow jersey (of world champion) later in the season. The Belgian could not have chosen a more direct route to Armstrong's heart. As for the job, it was his.

"The hiring of Johan was proof that Lance ran the team," says Vaughters. "I mean, Johan was the guy who gave Lance the confidence that he could win the Tour de France. He was the first person that could see Lance win the Tour, and I think Lance is tied to Johan because of that. But even with Johan on board, Lance still ran the team, just as he did back in 1998."

When it became clear that Armstrong wanted Bruyneel to take over as team manager, Weltz called Tyler Hamilton and asked if he would put in a good word for him with the team's bosses. Realizing Weltz was being ditched by one of the biggest bosses, Hamilton knew there was no point in trying to help him. O'Reilly didn't think Weltz had much right to complain. "I didn't help Johnny," she says. "Very few of us did. I didn't like him when I worked with him. He treated the staff badly and you soon get tired of that. Then there was Denis Gonzales, Johnny's mate, who was assistant director. Really, Denis was just a bit of a chamois-sniffer. That's what we called the girls who hung out around bike racing: groupies who just sucked up to the riders. There were a lot of men like that, too. You'd see them at the feed zone in races, and if they were coming in your direction, you would put your mobile phone to your ear and pretend to be speaking.

"The funny thing about Johnny was that as soon as he left the team, I discovered a different person. We got on great from the moment he left, and I really got to like him as a guy. He showed himself to be a really decent human being and I was sorry not to have stood up for him."

In December 1998, Postal's new directeur sportif, Johan Bruyneel, called O'Reilly, who was at her sister's home in Dublin, and formally offered her the job of head soigneur with the team. It was a significant promotion for O'Reilly, as it involved a good pay rise (from $30,000 to $36,000), and it came at a great time in the evolution of the team. After Lance Armstrong's fourth place in the Vuelta a España, success in the 1999 Tour de France was a possibility. O'Reilly knew he was mentally stronger than the Europeans. As head soigneur and Armstrong's massage therapist, she would have a position of real responsibility on the team. But before saying yes to Bruyneel, something had to be made clear.

"I would love the job, Johan, but I don't want to have anything to do with the medical program," she said. She would not inject riders with vitamins or anything else, she would not help to set up intravenous drips, and she would not carry substances that they themselves might not want to have in their possession. "Johan, I want to be head soigneur, but I don't know how to do the medical side and I don't want to know."

"Fine," he said. "Absolutely fine. I already know that."

Professional cycling has traditionally been wary of opening its doors to women. O'Reilly only had to look at the other teams to see how few of her sex were employed in the sport. In the beginning she had been pleased to get a job with a cycling team, and now she had one of the more prestigious jobs in professional cycling. Inside her head, a voice ridiculed her for refusing to get involved in the team's medical program and told her she was a phony. How could a head soigneur not be part of the program? But it wasn't a problem for Bruyneel and Armstrong. She actually felt they were pleased she wasn't involved and that it might have been part of the reason she was given the job. If O'Reilly didn't know the medical program, she couldn't get anyone in trouble by talking about it. And Bruyneel, in any case, believed the doctors should be running the program, not the soigneurs.

Above all, O'Reilly's promotion came about because she was good at her job. "I really liked Emma," says Marty Jemison, who rode for the team for five seasons, the same five years as O'Reilly. "She was a brilliant soigneur. A really hard worker, and you know, at times, she could be very demanding. She's got that fiery Irish personality, but she took care of us, she really took care of us. She gives a phenomenal massage, and when Emma was around, everything was impeccably clean and well organized. I often wondered did she, in the very back of her mind, hope that someone on the team would fancy her, but she was totally professional. Totally, totally professional. Was she the best I worked with? You bet she was."

Jonathan Vaughters is equally generous in his assessment. "Emma was always very good to me. She had a great sense of humor. She did her job very well. She was very demanding of those around her because her standards were high. She worked her rear end off and wanted others to do the same. No matter what it took, she got the job done. If she had to stay up until three in the morning doing water bottles, she stayed up. I got along great with Emma but pretty much everyone got along great with her. I mean, she was a young woman working in a very male environment but she handled that amazingly well. I imagine if the female soigneur was plain or unattractive, it might not have been that difficult. But Emma's a cute girl and she was the same age as the riders and she had the same energy. As far as I know, she was 100 percent professional and that's a big thing to say about someone in that environment."

Frankie Andreu is another former Postal rider with fond memories. "I liked Emma a lot and would regard her as a friend. She was very professional and very good at her job. Probably the best soigneur I worked with."

Perhaps the most glowing testimony of all came from the team's general manager, Mark Gorski. "Emma is the heart and soul of the team," he said in a July 1999 interview with the *International Herald*

Tribune. "She's so professional and has a wonderful influence on the other staff members and the riders. She brings a woman's sensitivity to guys' personalities, their differences in personality, all kinds of things that a male soigneur just can't bring."

At the end of the 1998 season, the personnel in the Postal team changed significantly. Bruyneel replaced Weltz as directeur sportif. Dr. Pedro Celaya left the team to join the Spanish ONCE team; another Spanish doctor, Luis Del Moral, replaced Celaya; and Jose Arenas, the master of the injection, left the team and his place was taken by his Spanish compatriot Jose Marti.

Within the team, Marti was known as "Pepe" and he was officially listed as a trainer. That puzzled a number of people because Marti didn't quite cut it as a physiologist or coach. At different times he would talk to riders about what training they should be doing, but technical conversations with him tended not to last too long. No one wanted to see him try to engage Vaughters in a discussion about training methods because that would have been embarrassing. Vaughters was very knowledgeable. So they looked at Pepe and tried to figure out what he did. For sure, he worked a lot with Del Moral, the doctor, but no one could figure out the nature of their relationship. But as the season progressed, and Pepe turned up at the team hotels with supplies from the pharmacy, they realized he was stocking up the doctor's medical cabinet. "The courier"—that's what they used to call Pepe Marti.

But he was not the only staff member used to transport products.

Chapter 10

CROSSING THE LINE

In early May 1999, Lance Armstrong finished a short training camp in the Pyrenees. He had gone there to reconnoiter the routes he would race in a Tour de France that was just two months away. A select group of people from the Postal team were invited to accompany him. There was directeur sportif Bruyneel, team doctor Luis Del Moral, mechanic Julien DeVriese, and Emma O'Reilly. By this time, they were all beginning to believe Armstrong would be hard to beat in the Tour de France, and even if O'Reilly was no longer a romantic when it came to professional cycling, she still enjoyed being part of the team's inner sanctum. As the group was breaking up, O'Reilly says she was asked to do something that she somewhat uneasily agreed to do. She related the story in an interview in 2003 and retold it under oath three years later. According to the masseuse, she was asked by Armstrong if she would drive down to Spain and collect a medical product from Del Moral at the team's base in Piles on the east coast. "Oh, okay," she said, "I'll do it this time." Not wanting to make such a long drive alone, O'Reilly intended to take her fiancé, Simon Lillistone, with her and mentioned this to Armstrong. She says Armstrong told her not to tell Lillistone, but, feeling she couldn't ask Lillistone along without telling him the purpose of the trip, she did not answer when Armstrong advised discretion.

After training camp ended, Bruyneel drove to his home in Spain in the U.S. Postal Service car O'Reilly had used to get to the Pyrenees. To return to Lillistone's apartment in Valras Plage, she hitched a ride with Julien DeVriese, and to make the journey to Piles, she hired a car in the French city of Béziers, using her U.S. Postal Service credit card. At a Citroën garage, she chose a navy-blue Xsara and wondered if Bruyneel had deliberately taken the team car so she would have to use a less conspicuous rental car. It was late Friday afternoon when she and Lillistone left the south of France. They traveled by autoroute and it took them five hours to get to Piles. It was nine-fifteen at night when they arrived and, exhausted, all they wanted was sleep. But Louise Donald, girlfriend of the Postal mechanic Geoff Brown, was keen to talk and they stayed up for a while chatting.

The U.S. Postal Service team rented two houses in Piles. One was for Brown and Donald, and they sublet a room to O'Reilly. Other staff members and a few of the riders used the second house. Each had a basement garage and the one under Brown and Donald's house was used by the soigneurs. The mechanics used the other one. The morning after she arrived, O'Reilly went for a run. With its warm, bright sunshine, Piles was one of her favorite places. A few turns and she was out of the town, between the rows of orange trees, and then on to the seafront. There she hung a left and ran about a mile up the coast before retracing her route back to the house. That Saturday afternoon, O'Reilly organized the trucks for forthcoming races. It was a pain but she liked working with Ryszard Kielpinski, the team's Polish soigneur.

They restocked the riders' clothing, replaced the food supplies, made a shopping list of what was needed, and then cleaned the inside of the trucks, which always took forever. While they were doing the trucks, Johan Bruyneel showed up. Both in her interview and in her sworn testimony, O'Reilly is definite about what happened next. "I was standing on the downhill entrance to Geoff's garage when Johan discreetly slipped a bottle of pills into my hand," she said. "He handed it to me like this, the bottle concealed in the palm of his hand. He just moved alongside me and I took it without anyone seeing. Johan would have been concerned about Louise being there because she talked a lot. Johan was very pleasant and I thought I must be doing him a big favor for him to behave like this. I didn't ask Lance what I'd be carrying because I didn't want to know. But I did know. If this was a legal product, we would have been able to buy it in France. Of course I was doing something wrong. Otherwise why would Lance have asked me not to tell Simon? And Johan was being too nice. Whatever was in that bottle, it wasn't paracetamol. The bottle containing the pills was no taller than three or four inches and it was round. Like a normal bottle of pills that you'd get from the doctor with any prescription. You could see the white tablets inside the brown plastic; there were maybe twenty-four tablets. I went into the house and put the bottle safely in my toilet bag."

On Sunday morning, the Postal crew all headed for the beach in Piles and then had lunch at a restaurant overlooking the sea. After eating, O'Reilly and Lillistone set off on the long journey back to France. It was dark by the time they reached the Spanish customs checkpoint, and a line of cars waited to be waved through. All the times she had crossed the border, O'Reilly had never waited in line. Now the waiting played on her nerves. She reminded herself that an unmarked rental car was not likely to be searched, but what if she were stopped and the pills were found? At that moment, she felt it was a mistake to have brought Lillistone along because he had nothing to do with the Postal team. If she was taken to a Spanish police station and allowed one telephone call, she decided it would be to Thom Weisel, who would know the best lawyers. She thought of Willy Voet, the Festina soigneur stopped on the French-Belgian border a year before, and how his first reaction was to protect the team and say the drugs were for his own use. Sorry, but she wasn't taking the rap for anyone. What struck her was the lunacy of what she had done and how so many soigneurs routinely did this kind of thing. "Glorified drug smugglers," she thought.

Then she edged her Citroën to the crossing and the Spanish customs official waved her through.

O'Reilly spent that Sunday night at Lillistone's apartment in Valras Plage and agreed to meet Armstrong in the parking lot of a McDonald's fast-food restaurant on the outskirts of Nice the following morning. She placed the bottle of pills in the pocket of the driver's door in readiness for the handover. The rendezvous was timed for eleven-thirty but it was almost midday when she got there. Armstrong normally didn't like to be kept waiting and en route O'Reilly called to apologize, but he was remarkably understanding. It was all right. She was to take her time. Arriving at the McDonald's parking lot, she parked to the right of Armstrong's navy-blue Passat Estate. The rider's wife at the time, Kristin, was in the passenger's seat dressed casually in a sweatshirt and shorts. O'Reilly thought that unusual for Kristin. O'Reilly testified that Armstrong then walked to her car and she handed him the brown bottle. She also recalled Kristin getting out of the car, but Kristin didn't witness the handover, which happened in a matter of seconds. The trip to Spain was never again mentioned.

Two weeks after the Pyrenean rendezvous, another select group of Postal riders and support staff headed for the Alps, where the team leader trained over the Alpine mountain routes he would tackle in the Tour de France. In this party were Armstrong and Bruyneel; Livingston and Hamilton; Del Moral; Bruyneel's wife, Christelle; and soigneurs Peter Van Boken and O'Reilly. On May 18, Dr. Michele Ferrari arrived in his camper van at Sestriere, the ski village in Italy where the team had set up camp. For some of the staff, it was a first time to see Ferrari, but the esteem in which he was held by Armstrong ensured he was warmly welcomed. That evening they ate at the Last Tango restaurant in Sestriere and it was clear how comfortable Armstrong and Ferrari were in each other's company. They sat together and chatted amiably through dinner. The more observant at the table would have picked up on the disaffection apparent in Dr. Del Moral's body language. He wanted to be more involved in Armstrong's preparation but had been kept at a distance. As he watched the rider and his Italian doctor shoot the breeze, Del Moral could see why he had been kept at a distance and it didn't improve his mood.

As remarkable as the chemistry between Ferrari and Armstrong was the secrecy of their relationship. U.S. Postal Service riders Jonathan Vaughters and Marty Jemison would not discover that Armstrong worked with Ferrari until after they left the team—in Vaughters's case, 2000; 2001 for Jemison. All of those who ate at the Last Tango understood the need for discretion and, more important, observed it on their return to the larger team group. Armstrong, Livingston, and Hamilton would go to the training camps without telling their teammates where they were going, and on their return they would not say where they had been or whom they had been with. Livingston had been working with Ferrari for at least three years, but Hamilton's situation was different. A strong climber, his importance to the team was obvious and explained his presence in Sestriere, but at first, he didn't have a formal working relationship with Ferrari. Within the team, people wondered if he didn't want to pay the doctor's exorbitant fees.

Although he was encouraged by Armstrong to work with Ferrari, Frankie Andreu said no. Andreu had been told by Armstrong and others that Ferrari knew everything about physiology and was a master at drawing up a training program, but Andreu preferred to continue in his own way. Of course, it wasn't just that. He had heard the rumors about Ferrari, he had seen what had been written in the press, and though he knew nothing for a fact, he just didn't like Ferrari's reputation. Some of the guys asked him why he didn't want to work with Michele, and he said it was more like he didn't want to be associated with the guy.

Like everyone in cycling, O'Reilly was aware of Michele Ferrari, but before that rendezvous in the Alps she had not seen him. No one told her Armstrong worked with him but she wasn't surprised he did. "I knew Ferrari's reputation in cycling was dirty. Everyone seemed to think he was the best, and he usually had the top riders going to him. When you're on the inside, you see things differently and make judgments that would not be logical in normal life. Riders who worked with Ferrari had to pay him out of their own money, whereas the team doctor was free. I thought the guys who went to Ferrari were prepared to make a proper investment in their careers and I respected them for that. Tyler didn't go to Ferrari and I thought it was because he was too tight with his money. He would listen to the things he [Ferrari] was doing for Lance and Kevin but he wasn't prepared to put his hand in his pocket. I didn't give too much thought to the fact that what Ferrari was doing was probably unethical and immoral."

O'Reilly would later hear from other staff members that Ferrari turned up at the team hotel during the 1999 Tour de France. Stefan DeVriese, the son of the mechanic Julien, got him a jacket from one of the motorbike riders on the race and Ferrari was delighted with that. Whenever the staff spoke about him, they referred to him by his code name of the time: McIlvenny.

On June 6, 1999, the Postal team competed in the Critérium du Dauphiné Libéré in the south of France. This is the second biggest stage race in France, and though it lasts just one week, it can be extremely tough. And, coming three weeks before the Tour de France, the race is considered ideal preparation for the Tour. In the 1999 race, the Postal team rode well, and after winning the key time-trial stage to the summit of Mont Ventoux, Vaughters took over the yellow jersey. With Armstrong's eyes fixed on the following month's Tour, he was happy to help Vaughters in this important but nevertheless preparatory race. Needless to say, he expected his teammate to hold on to the leader's jersey and record what would have been the U.S. Postal Service's greatest victory to that point.

In the penultimate stage, a difficult mountain race, the attack came from the Kazakh rider Alexandre Vinokourov, and as hard as

Vaughters tried, he could not keep up. Armstrong stayed by his teammate's side and tried to pace him back to the lead group but Vaughters didn't have the strength. It was a torturous experience for him; not only did he lose his lead and end up second to Vinokourov, he held back Armstrong who was very strong. Afterward, there was no sympathetic pat on the back from Armstrong, no acknowledgment that Vaughters had given it everything and that second place in the Dauphiné was a worthy achievement. The Postal leader didn't do sporting commiseration. As for Vaughters, he felt he'd let the team down: they had worked for him, placed their trust in him, and he had not delivered.

Others on the team thought Vaughters's keen intelligence wasn't much of an advantage in his career, if indeed it was an asset at all. He asked questions about what the doctors were doing and always wanted to know if what he was being offered was legal and whether it was necessary. It wasn't that he didn't recognize the need for recovery products but he wanted to know what they were and whether there might be any long-term effects. Questions such as these might be considered normal in civilian life, but in professional cycling they were looked upon as unnecessary, even unhelpful. Riders were expected to commit to their métier, to fully trust the doctor or soigneur, and to embrace the medical requirements. Vaughters couldn't surrender himself in that way, and in one sense, he stayed on the periphery of the Postal team. At the end of the 1999 season he left to join the French team, Crédit Agricole, and, surprisingly, found it was easier to be the rider he wanted to be with his new team.

Emma O'Reilly remembers the activity followed by the disappointment of Vaughters's week at the Critérium du Dauphiné Libéré. She liked Vaughters and so wanted him to win. But it is not her only memory from the race. Either from the Postal leader himself or through someone else, she learned Armstrong's hematocrit was forty-one, and even with her limited knowledge of these things, she knew that with a hematocrit that low, it was hard to be in any way competitive in cycling. During massage one evening, she just said, "Ah, that's terrible, forty-one. What are you going to do?" According to

O'Reilly, Armstrong looked up and said, "You know, Emma. What everybody does." She understood that to mean he would take r-EPO.

The relationship between a rider and his soigneur is normally close. Each evening they spend an hour or so in each other's company as the rider receives massage. The best soigneurs tend to the psychological needs of the rider as well, making him laugh when he needs to laugh, offering reassurance in moments of doubt, and knowing when it is better to be silent. Emma O'Reilly and Lance Armstrong were a good team, everyone on the U.S. Postal Service team could see that. She was an excellent masseuse and in an environment where others tended to genuflect before Armstrong, she treated him as a normal human being, which he seemed to like. They went to the 1999 Tour de France as a team within the team.

The race began on Saturday, July 3, and as is customary, the Tour's official medical examination of the participants took place on the eve of the start. Though it is a ceremonial event scheduled more for its publicity value than for medical reasons, the riders are obliged to turn up. Competitors know in advance that when they strip to the waist and have their lung capacity measured or their heart rates checked, photographers will be present. On this occasion that presented a problem. According to O'Reilly, they were on their way to the medical when Armstrong asked her if she had any makeup that would cover a bruise mark caused by injections to the outside of his upper arm. O'Reilly explained that what she used wouldn't be much good on his arm, but sensing he wanted this taken care of, she went to a pharmacy and bought a concealer-type product. He rubbed it over the bruise and they both laughed because O'Reilly didn't think it was very good but he thought it was okay.

Cyclists get legal vitamin injections and intravenous drips all the time, so why should Armstrong have been bothered about the bruise marks on his arm? A former Tour de France doctor who has devoted much time to the study of performance-enhancing drugs in sport, Jean-Pierre de Mondenard, says injections to the outer side of the upper arm are very specific. "Vaccinations, insulin, r-EPO, or growth hormone are injected there. All authorized substances—vitamins, iron, and substances to aid recovery—are injected into muscle in the buttock. An intramuscular injection can also be given in the thigh.

"Injections can be given only to a limited depth in the upper arm, so shallow that insulin needles have to be used. And it is easier to cause bruising there." Jerome Chiotti, a former mountain bike world champion, wrote a book called *De Son Plein Gre* (Of His Own Free Will) in which he admitted using banned performance-enhancing drugs at the height of his career. He says there is no mystery about injections into the upper arm. "That's where growth hormones, r-EPO, or corticoids may be injected," he said. "Personally, I injected r-EPO into the fold of the stomach or the upper thigh, but everyone has their own way of doing things. In my opinion, in that particular place [the upper arm], there's a 99 percent chance it was corticoids." Why could the bruise not be caused by legal substances to aid recovery? "It's no use injecting them into the arm," Chiotti said. "Iron and vitamins are taken either by intravenous drip or injected intramuscular into the buttock. The arm is not suitable and in any case, it would be very painful." Willy Voet, the Festina soigneur with twenty years' experience of doping riders, offered his assessment. "We injected growth hormone, r-EPO, corticoids, or even amphetamines into the upper arm. As for other substances, iron and vitamins, we injected them into the buttock." Could the bruise mark not have been caused by glucose injections? "Injections of glucose are always intravenous, never subcutaneous, and into a vein on the inside of the lower arm," said Voet.

O'Reilly's concealer worked and Armstrong's presence at the medical examination passed without a hitch. The next day he rode the prologue at Le Puy du Fou and blitzed his rivals with a scintillating performance. "When he won, we all dropped dead," she says. "We kind of expected him to do something but we weren't thinking he could win the prologue. But he did and it was great, fabulous. Even Mark [Gorski] was delighted. I stood at the finish line, he came in, I heard the time being called out. Excitement everywhere and I'm thinking, 'What?' I spoke to Kristin [Armstrong's wife at the time] and said, 'We've won, we've won.' It was a lovely moment. On the following night there was champagne. I was having my shower when Mark knocked on my door, and I thought, 'Can a woman not take a bloody shower without being disturbed?' He said, 'Emma, can you hurry down? We have a bottle of champagne and we don't want to open it until you arrive.' That was thoughtful of Mark and it showed there was a bit of camaraderie in the team."

On the evening before the 1999 Tour de France began at Le Puy du Fou in the Vendée region, the directeurs sportifs of the twenty-one competing teams got together for their customary prerace meal. Also at this meeting was the president of the Union Cycliste Internationale (UCI), Hein Verbruggen. His message to the team directors was important: from now on the UCI would be testing for banned corticoids. Corticosteroids or corticoids are hormones produced by the body or manufactured in synthetic form; all are steroids with similar chemical structures but different physiological effects. These drugs had been proscribed in 1978 but, undetectable, they were widely abused by cyclists and athletes in many other sports for the following twenty years. Shortly before the start of the '99 Tour, the French National Laboratory for the Detection of Doping (Laboratoire National de Dépistage du Dopage, or LNDD) at Châtenay-Malabry came up with a drug test that identified the presence of corticoids in urine, and the UCI adopted it.

When, asked the team managers, would this test be implemented? "From tomorrow's prologue," said Verbruggen. To many of the team directors present, this came as a shock. Within professional cycling, the abuse of corticoids was endemic, and team officials would have expected some warning before having a new test foisted upon their riders. No one was quite certain whether riders testing positive for corticoids would actually be sanctioned. Jacques de Ceaurriz, head of the French laboratory, saw the tests as not necessarily leading to sanctions. "This is more of a preventative measure," he said. "It will give us another picture of the general health of the cyclists. Riders found to have used corticoids may not be deemed positive, especially if they are able to provide evidence of medical treatment." Leon Schattenberg, head of the UCI's medical commission, took a different line. "If there is a positive test, we will contact the rider concerned. He will be entitled to a second analysis, and if that confirms the result of the first, he will be punished." Schattenberg also said he believed corticoids concerned "a very small number of riders." That was not a view shared by De Ceaurriz, based on his testing of samples at the previous year's Tour de France. "If those using corticoids in that race had been sanctioned," he said, "the Tour would not have been finished due to a lack of competitors."

After the prologue to the 1999 Tour ended, four riders submitted urine samples at doping control. One of the four, the sample taken from the Danish rider Bo Hamburger, tested positive for corticoids. The following day's race was from Montaigu to Challans in western France and Lance Armstrong successfully defended the race leader's jersey he'd won in the prologue. As race leader, he had to report again to doping control. It was July 4, Independence Day in his home country, and it was also the day he tested positive. Traces of the synthetic corticoid triamcinolone acetonide were found in his urine. According to the UCI's list of banned substances, the use of corticoids was controlled as follows:

Figure 111 Classes of drugs subject to certain restrictions

> C: The use of corticoids is prohibited, except when used for topical application (auricular, ophthalmological or dermatological), inhalations (asthma and allergic rhinitis) and local or intra-articular injections. Such forms of utilisation are to be proved by the rider with a medical prescription.

Proved by the rider with a medical prescription? On Armstrong's doping form at Challans the word *néant* (none) was written alongside the column marked *Medicament Pris* (medications taken). Over two weeks would pass before the story of Armstrong's positive test would be leaked to the press. During that time there were rumors of riders testing positive for corticoids but nothing more. Another French newspaper, *l'Humanité*, suggested the number was as high as "twenty to thirty cases." On July 19, a rest day on the Tour, Armstrong gave a press conference in the southwest city of Tarbes, during which he claimed never to have taken corticoids and never to have needed a medical exemption for any banned product. This was a reaffirmation of what he'd said in an interview with a journalist from the French sports daily *L'Equipe* eleven days before. On the evening of the rest day (July 19), Benoît Hopquin, a journalist from the French daily newspaper *Le Monde,* received a tip-off indicating that Armstrong had tested positive following his July 4 test. Two other *Le Monde* journalists, Yves Bordenave and Philippe Le Coeur, worked with Hopquin to check out the story but couldn't get it confirmed. The next morning, another *Le Monde* journalist met the original source of the tip-off and was allowed to see the relevant doping report form for Armstrong's July 4 test.

Emma O'Reilly was already aware of Armstrong's positive from a chance conversation with Kevin Livingston twelve days before. That evening, Livingston, one of the three riders she looked after, was late for his massage because he had been selected for a random drug test. O'Reilly says Livingston told her that an anti-doping official had told them about the corticoid positive, and his (Livingston's) feeling was that someone in authority wanted to warn the Postal team about the result. That didn't surprise O'Reilly, who felt there was collusion between the authorities and the teams on the question of anti-doping. She also felt that after the scandal of the previous year, no one wanted a doping controversy in the '99 race.

Because she was not involved in the team's medical program, O'Reilly could remain detached from such controversies. She heard the talk within the team but, mostly, she stayed out of the conversations. More than a week after Livingston's call to a random drug test, O'Reilly heard that *Le Monde*'s reporters were on to the story. Within the team, there was talk and a sense that something was about to break.

After the rest day, the race resumed with a stage through the Pyrenees finishing at the ski station in Piau-Engaly. By then Armstrong was back in the yellow jersey and his grip on the race was viselike. He traveled by helicopter from the summit finish that evening, flying high over the snaking line of congested traffic, and arrived at the team hotel long before his teammates. O'Reilly was part of that barely moving convoy as it inched its way down the mountain, but she was consoled by the thought of a lighter workload that evening. "One of my three, Jonathan Vaughters, had already crashed out of the race and so I was down to just Lance and Kevin. Because he went by helicopter, Lance got to the hotel about two hours before me and I presumed one of the other soigneurs would have given him his massage. So I thought 'Well, I'll have just Kevin this evening, won't be too bad.' But when I got to the hotel, Lance was there, sitting on his bed, waiting for me. I thought that was kind of cute."

That same Tuesday night, *Le Monde*'s journalists tried to get a reaction from the U.S. Postal Service team but were told the team would wait to see what the UCI had to say. At around ten o'clock that night, one of the *Le Monde* journalists spoke with Dan Osipow, Postal's media officer. Still, he was unprepared to comment. The journalist specifically asked if Armstrong had used a cream that could have caused a positive test, and Osipow said he could not respond to that. *Le Monde*'s interest in the story put pressure on the team to explain the presence of the corticoid in Armstrong's urine. According to Emma O'Reilly, Postal's response was decided during the delayed massage she gave to Armstrong that evening.

"At one stage, two of the team officials were in the room with Lance. They were all talking. 'What are we going to do, what are we going to do? Let's keep this quiet, let's stick together. Let's not panic. Let's all leave here with the same story.' There was a real sense that the shit was about to hit the fan and they had to come up with an explanation. From that discussion came the saddle sore story, the corticoidal cream to treat it, and a backdated medical prescription. I knew about the cortisone because Lance told me he had taken it before or during the Route du Sud the previous month, and because there were no problems there, he presumed he would be all right in the Tour de France. I don't remember any talk of Lance having saddle sore at the start of the Tour, but anyway, he told me it wasn't the cream. Later that night there was a mad scramble to get Luis [Del Moral, the team doctor] to write the medical prescription."

On July 22 the UCI issued a communiqué stating Armstrong used the ointment Cemalyt, which contains the corticoid triamcinolone, to treat a skin allergy. The UCI went on to say it had seen the medical prescription for this ointment but, significantly, the communiqué did not specify when. Neither did it say whether the rider had declared his use of Cemalyt on his doping form, as he was required to do. Toward the end of its statement, the UCI warned journalists about jumping to conclusions in doping cases:

> We should like to ask all press representatives to be aware of the complexity of issues and the related aspects of the rules and the law before producing their publications. This will allow considerations of a rather superficial, not to say unfounded, nature to be avoided.

At a press conference a few days later, Armstrong went on the attack against *Le Monde,* calling it "the gutter press." Hopquin asked why he had twice stated that he had no medical exemptions for banned products when he was now claiming he had, to which Armstrong replied: "Mr. *Le Monde,* are you calling me a liar or a doper?" Quite a number of journalists laughed when Armstrong asked this question, the mirth seeming to express their support for the rider. In a room full of journalists, the questioning reporter from *Le Monde* was on his own. No one asked a follow-up question and the corticoid affair ended.

O'Reilly has recalled the short conversation she had with Armstrong as the U.S. Postal Service team officials left the massage room on that fateful evening. He said, "Now, Emma, you know enough to bring me down."

No one who followed the 1999 Tour de France will forget Lance Armstrong's performance. He started by winning the prologue on the very first day, then he took control of the race after a stunning victory in the first time trial at Metz, and in the mountains he towered over his rivals. On July 14, Bastille Day in France, he rode an imperious race to the summit of Alpe d'Huez, the most renowned of the Alpine peaks. Though she remembers Armstrong's total control on that afternoon, Emma O'Reilly's more vivid memory of the Alpe is a conversation she had with Christi Anderson, the Eurosport commentator and wife of the former Motorola cyclist Phil Anderson. O'Reilly noticed a handsome Rolex watch on Anderson's wrist, not big enough to be a man's, not small enough for a lady. But that was exactly what O'Reilly liked about it. "Ah, that's a lovely watch, Christi. Give us a closer look," she said.

"Emma, these cost four thousand dollars. Phil bought it for me."

If she had used a wire brush, Christi Anderson couldn't have done a better job of rubbing O'Reilly the wrong way. "You know, love," O'Reilly said, bristling, "if I want to spend four grand on a watch, I will. Don't turn your nose up at me just because I'm a staff person."

At the time, O'Reilly was engaged to Simon, and as much as she wanted a nice watch, she couldn't justify it. While giving Armstrong his massage that evening, she recounted the conversation with Anderson. "The cheek of her. Who does she think she is, thinking I couldn't buy a Rolex if I wanted to? But that kind of money on a watch is a bit stupid, isn't it? If I wanted to, though, I could." Armstrong listened sympathetically and agreed Christi Anderson had some nerve.

Two weeks later the Tour ended in Paris and O'Reilly was hellishly busy, unpacking the Fiat cars loaned to the teams for the duration of the race, and making sure the right pieces of luggage went into the correct team trucks. She was tired, grumpy, and a long way from sorting out the chaos when, from on high, word came that she and Julien DeVriese had to go to the Ritz-Carlton, where Armstrong wanted to see them. With so much still to be done, O'Reilly didn't need the distraction. She and DeVriese were physically wrecked, emotionally drained, wearing shorts that needed a washing machine, and Trek shirts that O'Reilly felt would be ideal for serving gasoline. But when the leader calls . . .

They got to the Ritz-Carlton a little after Armstrong finished his postrace press conference and were told by the lady at reception that Monsieur Armstrong left word he was not to be disturbed. Having made the effort to get to the hotel and curious why Armstrong wished to see them, O'Reilly and DeVriese were not about to give up.

"Call his room, tell him we're here. We work for the team," said O'Reilly.

"I'm sorry, I can't do that."

"Can we speak to the manager? This is very important."

The manager came but insisted nothing could be done. "Mr. Armstrong said he wasn't to be disturbed."

"I don't care," said O'Reilly. "You just tell him Emma and Julien are down here and he'd better get a move on because we're not in good humor."

"Look, if you're his friends, can't you ring him on his mobile phone?"

"I can't because I've got his bloody mobile phone here in my hand."

Understanding this storm was not going to abate, the manager relented. Beckoning a porter, he told him to go up to the room, knock on the door, and see if Monsieur Armstrong wished to speak with these people. O'Reilly and DeVriese followed the porter, who, on arrival at Armstrong's door, knocked in the gentlest manner. "I don't want to be disturbed," Armstrong shouted from within. At that moment O'Reilly remembered she had Kristin Armstrong's mobile phone number and she called it. Armstrong himself answered.

"We're outside your door, you moron," she said.

"Okay," he replied, and two seconds later the door opened and everyone was smiling, especially the porter. Inside, Kristin was cutting her husband's hair. "Just let Kristin finish this," Armstrong said.

"We'll talk then." Bedroom, bathroom, and spacious living room—the Tour de France champion was already living like a king. While waiting for the haircutting to end, O'Reilly picked up the Ritz-Carlton pen and said to DeVriese they should take the soaps and shampoos as well, just so they could pretend they stayed.

His hair cut, Armstrong stood up and produced two magnificent Rolex watches. The presentation was low-key but intimate; his aim was to show gratitude to two people who had been very important members of the team. Soigneurs and mechanics don't normally receive Rolex watches from their bosses, and the unexpectedness of the gift made it all the sweeter for O'Reilly and DeVriese. Julien laughed as he and Emma left the hotel. "You'd know Kristin bought them," he said. "He would never have spent that much money on us." But they both understood it was Armstrong who had asked his wife to buy them.

Back in England, Emma got the watch insured. It was valued at four thousand dollars. She noticed the date on the guarantee—July 16. The watches were bought two days after that fraught conversation with Christi Anderson at Alpe d'Huez.

Chapter 11

PLUS ÇA CHANGE, PLUS C'EST LA MÊME CHOSE

(The More Things Change, the More They Stay the Same)

Bruno Roussel is a Breton, and like so many from Brittany he exudes a straightforwardness not routinely encountered in other parts of France. For his troubles, Roussel was directeur sportif of the Festina team in July 1998, when the team's soigneur Willy Voet was stopped on a nondescript road just two hundred yards from the Belgian border. It was early morning when a customs officer signaled Voet to pull over, and what happened next would soon be enshrined in cycling's Grand Hall of Infamy. It may be forgotten what bike Marco Pantani rode when winning the '98 Tour de France, but it is widely known that Voet drove a Fiat. It is also widely known that it was packed with banned performance-enhancing drugs, and the French judicial inquiry that followed exposed professional cycling as drug-riddled and morally bankrupt. As the boss of a team with a systemic doping program, Roussel spent a lot of time helping police with their inquiries. He told them all he knew, which was a lot, and gently tried to pry from them the one thing he didn't know.

What Roussel couldn't figure out was how the customs officers, five of them, happened to be on a small road at Neuville-en-Ferrain in northern France at the very second Willy Voet was passing their way on that Wednesday morning. To begin with, Voet should have been traveling on the autoroute but, on a last-second hunch, turned onto the smaller road to lessen his chances of being stopped. Then, too, his was the very first car that was pulled in that morning, and a small army of customs officers was present to deal with the case—at six A.M. And it all took place when France was transfixed by another sport, hosting the soccer World Cup and waiting for its team to play Croatia in the semifinal at the Stade de France in Paris that very evening. Yet there the officers were on that small road, fortuitously positioned to uncover one of sport's greatest scandals. Except that Bruno Roussel just didn't buy the *fortuitous* bit. From early on in the case, Roussel felt Voet was followed from his home in the south of France the previous day and tracked as he diverted into Belgium to visit the home of the Festina team doctor, Eric Ryckaert. When Voet switched onto the smaller road before the French-Belgian frontier, Roussel figured a phone call was made to the customs people on the border.

While in custody, Roussel became friendly with police commissioner Romuald Muller, who was part of the team investigating the use of illegal drugs in cycling. Muller appreciated Roussel's honesty and their relationship progressed to the point where the police officer agreed to the directeur sportif's request that the Festina team be allowed to continue in the Tour de France. Roussel's logic was that if the majority of riders in the Tour were doping, it was wrong to exclude racers from one team. Muller accepted that and from the police's point of view, the Festina riders were okay to continue in the race. The plan was thwarted by the Tour de France organization, which could not tolerate *known* dopers in their race.

After Roussel was released from custody, he and Muller remained friends, and when they happened to meet at the conclusion of the Festina trial at Lille two years later, it was natural that they should go to lunch together. They ate in a modest restaurant but the conversation was lively until, finally, Roussel got to say what had been on his mind for so long.

"It wasn't an accident that Willy was stopped by the customs?"

With a nod of his head, Muller indicated it wasn't.

"Okay, so who tipped you off?" asked Roussel.

"I can't tell you that."

"I understand that, so I will mention a name, and if the name is correct, you just smile. Roger Legeay?"

According to Roussel, Muller smiled.

At the time of Voet's arrest, Roger Legeay was directeur sportif of the rival GAN team. Roussel knows this isn't proof that Legeay was the one who did his civic duty and reported the wrongdoing to the police, but he nevertheless believes that's what happened.

Roussel found it easy to be a good witness in the Festina affair. He had not been a professional cyclist and hadn't sworn to follow the sport's law of silence. One of the first things he said to the police when arrested was that he was opposed to doping—a remarkable admission for a man who organized a sophisticated doping program for his own team, but it was true. Roussel accepted the need for a doping program because rival teams had programs and, more important, most of his riders were using dangerous products without medical supervision. He also felt the UCI didn't have the will or the resources to deal effectively with the problem, and what other option was there? He chose to take control and allow Dr. Ryckaert to supervise what his riders were already doing.

Festina's program was not compulsory and if a rider didn't want to dope, that was his right. According to Festina's trainer, Antoine Vayer, the overwhelming majority of the riders doped. "We had twenty-four riders in the team. It's been written that three didn't dope and the rest did. The truth is that only one, Christophe Bassons, was clean. Most of them did all of the drugs, and a small number used just corticoids." The twenty-two-year-old Bassons, who joined the team in April 1996, impressed Vayer. "He's got some engine, this kid," the trainer told Roussel, after conducting initial physiological tests on Bassons. "In terms of his physical capacities, he's got the same potential as [five-time Tour de France winner] Bernard Hinault had." No one who knew Bassons had any doubts about how far he could go. After all, he hadn't started racing until he was seventeen, and three years later he was French time-trial champion and had won the world military championships during his year of national service. Robert Vidal, president of the Union Velocipedique of Mazamet where Bassons started, had also helped another son of Mazamet, Laurent Jalabert. "This one has the same ability as Jalabert. Remember his name: Bassons."

Bassons's talent might have put him into a small group of elite athletes, but his attitude to performance-enhancing drugs made him a freak. Before putting his signature to the contract with Festina, he brought up the issue of doping with Roussel. "I know there is doping in cycling," he said, "but there is no question of my getting involved in that."

"In any case," replied Roussel, "we have a rule that young pros are not allowed to do anything like that for the first two years. But you should know, after that, it is possible. If one day you change your mind, we have two doctors on our team who can help."

"Possible or not, I will refuse."

"I am glad to hear that," said Roussel. "I find it reassuring."

"I can win without doping," said Bassons.

"Surely, surely," said his boss.

Though he was young and without experience, Bassons wasn't so much offering an assessment as his absolute conviction. Though he came from a modest working-class background—his father was a construction worker—young Christophe worked hard at school and afterward got a diploma at the University of Toulouse. His parents instilled a set of moral values and he had the confidence—perhaps even the arrogance—to believe that he could remain true to those values in the dope-ridden world of professional cycling, and still be successful. To better understand the young man's idealism, consider him at the age of eighteen, winning race after race in just his second year of competition and then being sidelined by a knee injury. His doctor administered an injection of the corticoid Kenacourt into the troublesome joint. On his first training ride back, Bassons realized that not only had the pain from his knee disappeared but also he felt stronger and more dynamic than before the injury.

The same feeling of enhanced power was there on his next training ride and Bassons suspected it was the effect of the Kenacourt injection. Though his use of the drug did not violate anti-doping law, Bassons felt it was wrong to gain a competitive advantage from a medication. He refused to race for the following two weeks and resumed only when the effects of the Kenacourt disappeared. How his new teammates at Festina would have laughed if Bassons had been foolish enough to tell them that story. Not to worry, there would be other reasons to smile at the ways of the rider they came to call *Monsieur Propre*—Mr. Clean.

In May 1996, a couple of weeks after joining the Festina team, Bassons went with a number of his teammates on a reconnaissance trip to the Alps to prepare for that summer's Tour de France. Festina's biggest names—Richard Virenque, Laurent Dufaux, Laurent Brochard, and Christophe Moreau—were all there and on the second day they reconnoitered the route to the Italian ski station at Sestriere that would be a decisive stage in the Tour two months later. Though things started off that morning at a relaxed tempo, the speed increased as legs loosened and competitive juices began to flow. They knew where the intermediate sprints would be in the Tour and contested them fiercely, the winner accepting the applause of an imagined crowd. As the journey proceeded, the talking and the joking decreased until all you could hear was the staccato breathing on the slopes of the long climb to Sestriere. Toward the last part of the ascent, Bassons moved to the front of the group, taking on the responsibility of the faithful équipier as he tried to prove himself to Virenque and the others.

In the lead, Bassons pushed a little harder, fully expecting his teammates to counterattack. To his astonishment, they started to fall away until there was only Virenque at his shoulder. Virenque accelerated past him on the last ramps of the climb and, as should happen, the team leader got to the hotel first. Bassons arrived second, the others sometime later. "He's not bad," Virenque said to Roussel. "He can be useful to me; you've got to try and get him into races with me." A week or so later, Bassons rode his first important race for Festina, the six-day Grand Prix du Midi Libre, and after that the eight-day Critérium du Dauphiné Libéré. Though they were preparatory races for the Tour de France, they were tough and Bassons suffered terribly in the mountains. He could have accepted the physical battering if his Festina teammates had suffered similarly, but they didn't. Although they had been left far behind on the climb to Sestriere, Dufaux, Moreau, and Brochard were comfortably stronger than Bassons in both subsequent races. What the innocent one did not realize was that all three, as well as Virenque and other Festina riders, had been given injections of r-EPO, growth hormones, and other drugs by Voet after the Alpine training camp. They were different men in those two races.

During those two weeks of racing, Bassons suffered from "passive doping" and would continue to do so until retiring from professional cycling in 2002 at the age of twenty-eight. Passive doping works like passive smoking in that the nondoper, in the company of dopers, suffers damage to his health. It happens simply. Those who use products such as r-EPO, human growth hormone, and anabolic steroids race at a speed far higher than would be possible without drugs. In attempting to compete with the dopers, the clean rider pushes his body beyond its natural limits, and though he might keep up one day, it is not possible to sustain that level of effort over a long period. Refusing to believe doping makes such a great difference, the clean rider forces his body beyond natural boundaries, which often leaves him physically wasted and in need of a long rest.

Passive doping can be dangerous. At the weeklong Italian Tirreno-Adriatico race in 1997, Bassons spent himself keeping up with the r-EPO-fueled peloton; day after day, his body was depleted. Dr. Ryckaert measured his hematocrit at 37.5 percent and told him he was borderline anemic, but Bassons would not accept an r-EPO injection. On the last day, virtually out of his mind with pain and exhaustion and no longer in control of his bike, he hit a steel trash bin at the side of the road. He got badly cut up but it was the fear of what might have happened that most troubled him. Ryckaert was sympathetic but pragmatic. "You have potential but you will get nothing out of this sport if you continue to refuse our help. You have to raise your hematocrit." The doctor's offer of help was delivered in classic cycling-speak: no explicit mention of doping and no reference to any products.

The psychological toll on Bassons was another matter, as the effects of passive doping drained his morale. There were times when he wanted to be one of the boys. He thought about his parents; they didn't want him to dope, but he felt that if he was successful, they would be proud and probably wouldn't ask questions. Pascale, his girlfriend, was different. They had discussed it and she'd said that no matter what, he must never dope. Never, ever dope, she said. He loved her for seeing it like that.

Through the years when doping was rampant, Christophe Bassons refused to cheat and refused to give up. Within the peloton he became a figure of fun; an example to everyone of what happens to those who don't dope. No results, no money, no recognition, and, ultimately, no respect. Of course, *Monsieur Propre* had his uses. When the doping officials came to the hotel early in the morning to take blood samples, the strategy was always the same: *Get Bassons to go first, tell him to take his time, to stall them as long as he can.* Up in their bedrooms, his teammates diluted their blood and lowered their hematocrits to "normal" levels. There were times, too, when Roussel grew impatient with Bassons's inflexible ethics. Results were the team's lifeblood and Bassons contributed little. No matter how much he admired the rider's saintly character, the directeur sportif was part of a results-based business, and they all had their responsibilities.

Frankie Andreu, Lance Armstrong, and George Hincapie in Milan before the start of the 1999 Milan–San Remo classic. Hincapie wears the stars-and-stripes jersey as the U.S. champion.

"How do you like them fuckin' apples!" said Armstrong to Johan Bruyneel and Thom Weisel as he won the Alpine stage to Sestriere in the 1999 Tour de France. The apple theme was continued at the team celebration at the Musée d'Orsay in Paris on the night the race ended: from left, Peter Meinert-Nielsen, Frankie Andreu, Tyler Hamilton, Pascal Derame, George Hincapie, Lance Armstrong, Christian Vandevelde, and Kevin Livingston.

Monsieur Propre (Mr. Clean), Christophe Bassons, a drug-free rider who struggled to succeed.

Dr. Max Testa tried to convince the Motorola riders that doping wasn't necessary.

Celebrating victory in the 1999 Tour de France outside Chez Wayne's bar in Nice—from left, George Hincapie, Lance Armstrong, Frankie Andreu, and Kevin Livingston.

Frankie Andreu observes as his wife, Betsy, and Lance Armstrong prepare dinner.

As directeur sportif, Johan Bruyneel made an impressive start to his career. Seven attempts and seven victories in the Tour de France—he had good reason to stick close to Armstrong.

Marty Jemison rode for the U.S. Postal Service team from 1996 to 2000.

Andy Hampsten could soar like an eagle.

Dr. Prentice Steffen worked for the U.S. Postal Service team from 1993 to 1996 before a controversial parting of the ways. *(Beth Seliga)*

Greg LeMond and Bernard Hinault on the epic Alpe d'Huez stage of the 1986 Tour de France

At times, the press conferences at the 1999 Tour de France seemed as tough as the race itself.

Emma O'Reilly, left, soigneur with the U.S. Postal Service team from 1996 to 2000, shown here in 1999 with Kevin Livingston, Becky Burnett (soon to be Becky Livingston), Frankie Andreu, and Kristin Armstrong, who was then Lance's wife.

The U.S. Postal Service team's class of 2000: Armstrong and Andreu lead their well wrapped-up teammates on a training ride.

Jonathan Vaughters, a talented climber who never found a place in the inner sanctum of the U.S. Postal Service team.

Andreu, the équipier who wanted to be able to do his job.

Stephanie McIlvain, who visited Armstrong at Indiana University Hospital when he was being treated for testicular cancer.

Tyler Hamilton won a gold medal at the Athens Olympics in 2004 but tested positive a month later and was subsequently given a two-year doping ban.

Stephen Swart now rides his bike purely for the love of it. *(Jackson Foster)*

For Floyd Landis, the real race began after he reached the Champs-Elysées at the end of the 2006 Tour de France.

When Willy Voet was imprisoned, it seemed certain that cycling would slow down. It might have been tough luck for Festina to be the team caught in flagrante but it was an opportunity for a sport that had been careening out of control. For Christophe Bassons, it meant renewed hope and the possibility of discovering his worth as a cyclist. At the end of 1998, he joined another French team, La Française des Jeux, and couldn't wait for the 1999 season to begin.

Through the winter of 1998 people spoke of a new beginning and recognized the need for a cleaner, more ethical sport. Determined to put in place the controls that would underpin the new age, the French Cycling Federation (FFC) set up a program of longitudinal testing for its riders. Each professional was obliged to show up at a designated laboratory four times a season, and from the results the federation would ascertain normal blood values for each cyclist. If any value deviated significantly from a rider's norm, his team doctor was notified and the team was expected to withdraw the rider from competition until the suspicious value returned to normal. Although this new approach to testing was not welcomed by the riders, they accepted it; only the two most successful, Laurent Jalabert and Richard Virenque, spoke against it.

A key concern for the French riders was to see if foreign riders were also committing to new ethics. The answer came at the first important race of the season, the weeklong stage race from Paris to Nice, which was won by the Dutch rider Michael Boogerd. It wasn't so much Boogerd winning that turned heads but the fact that three of his Rabobank teammates placed second, fifth, and sixth. Four riders in the top six signified the kind of dominance no one expected from any team in Paris-Nice—especially not from Rabobank, who weren't that good. What bothered their French rivals was the ridiculous ease with which Rabobank controlled the race. At no point were they stretched. Through the eyes of the home riders, this was the worst of all possible worlds: they had changed while their rivals hadn't.

A week later, one of the senior French riders, Jean-Cyril Robin, broke the sport's *omertà* on doping to speak with a journalist from the daily newspaper *Ouest-France.* "We are actually at an impasse," he said, "because there are still many who dope. Cycling is now racing at two speeds [one for those who dope and another for those who are clean]. The police need to come back and search the team trucks. I believe that is necessary. The riders who are still doping need to be punished and that needs to be done in public. We must never again have the law of silence." Robin referred specifically to the manner of Rabobank's domination in Paris-Nice, and such was his frankness that it seemed important debates about doping could now take place in public. If so, it would not be for long. Even though the FFC president, Daniel Baal, complimented Robin for having the courage to say what needed to be said, UCI president Hein Verbruggen wrote to the rider and reprimanded him for the comments. Privately Verbruggen told Robin that because he had criticized a sponsor, Rabobank, the UCI was obliged to react even if he agreed with the substance of what the rider said.

Only one other person criticized Robin, and that happened a week later at the Circuit de la Sarthe, a small stage race in western France. "It was actually the first stage in Sille-le-Guillaume," said Robin. "We had just set out on a road through a forest when Lance Armstrong cycled up beside me. He told me I shouldn't have said that, not publicly. He wasn't aggressive. His tone was neutral but firm."

During the month of April, the second round of FFC blood tests was implemented and sixty-seven of the riders examined had excessive levels of iron. Christophe Bassons was the only rider with an iron deficiency. In the mistaken view that it increases red cells, cyclists have traditionally used excessive amounts of iron. As more sophisticated medical protocols developed, riders realized the foolishness of this. But then, ironically, r-EPO was introduced—and for it to be effective, iron supplementation was necessary. From their first blood tests at the 1996 Tour de Suisse, the UCI knew riders were using far too much iron, and while the organization could not determine the reasons for this, it presumed two factors were involved: the traditional belief in the importance of iron and, second, the need to fuel new red cells produced by r-EPO. What the governing body also knew was that excessive iron was a serious risk to health. Though iron is vital for cellular life and required in numerous metabolic processes, it is highly toxic when not maintained within cells or bound to proteins. Too much iron leads to "free iron" and the production of tissue-damaging free radicals. The long-term effects of iron excess are not certain but it is known that it can lead to irreversible organ dysfunction.

In 1999 the UCI discovered 45 percent of riders showed above-average levels of iron, and 25 percent were way above average. Even if some of the excess was explained by the traditional approach, the UCI also knew r-EPO was a major contributory factor. Because iron excess was potentially injurious to health, the authorities could have introduced a rule insisting iron levels had to be normal before riders were allowed to compete. Such a rule would have forced the riders to take greater care of their health and it would have diminished the use of r-EPO. The UCI chose not to implement any such rule.

It was during the spring classics of 1999 that Bassons realized nothing had changed. For those prestigious races, the peloton raced at the same brutal pace of the previous five years. Bassons loved the Paris to Roubaix classic, and believing he could do well in the race, he gave himself, body and soul, to the effort until there was nothing more to give. He crossed the finish line in fortieth place, thirteen minutes after the winner, utterly wasted but also elated. It was his third attempt and the first time he had finished this race, so punishing it is known as "the hell of the north." And Bassons had raced it clean.

After the dirt of the route was washed away and the thrill of reaching Roubaix subsided, Bassons thought again about the hopelessness of his predicament. His new team, La Française des Jeux, and its directeur sportif, Marc Madiot, a former rider, appeared open to the new age, but Bassons didn't see how it was possible. Yes, many of his teammates were trying to ride clean, but without enthusiasm and with a constant anxiety about what other teams were doing, especially non-French teams. Bassons wondered if his teammates weren't psychologically dependent on doping.

The team also hired a number of young professionals, riders just out of amateur ranks, and these new recruits talked about doping products with an expertise that astonished Bassons. Only three years had passed since he'd ridden with the amateurs, and in that short time, the gangrene had clearly spread to that branch of the sport. And then there was the moment in Paris-Roubaix when Bassons rode a section of cobbles a few places behind a teammate who crashed. As his teammate hit the ground, a small tube fell from the pocket of his jersey and opened, spilling a couple of pills that Bassons could tell were caffeine pills. At the time, it was funny to see the pills on the roadside, but when he thought about it afterward, it didn't seem so amusing. Caffeine is not illegal but the rider who set out in Paris-Roubaix with a tube of pills still raced with the doper's mentality.

For so long, mental toughness and Pascale's ardent support saw Bassons through. It didn't matter what happened in the peloton or how often he was left behind, he continued to believe it was possible to survive without doping. He prepared for the 1999 Critérium du Dauphiné Libéré and Tour de France by training intensely and spending a week in a hypoxic chamber that replicates conditions at high altitude by sucking oxygen from the air. The body's response is to produce new red blood cells to cope with the diminished supply of oxygen. After his week in the hypoxic tent, Bassons's hematocrit did not rise by one point.

In the Critérium du Dauphiné Libéré, Bassons enjoyed a rare moment of success by breaking away early on the final stage, establishing a good lead, and holding on all the way to the finish. His breakaway lasted seventy-five miles and it was his best win as a professional. He knows, though, it was the final stage of a race already decided the previous day when the Kazakhstan rider Alexandre Vinokourov won the yellow jersey from U.S. Postal Service's Jonathan Vaughters; as Bassons was well down on general classification, no one was too bothered when he made his escape. He couldn't threaten Vinokourov's lead, nor even take Vaughters's second place.

About the only one who thought to chase him was Lance Armstrong. "Lance thought we should chase," says Vaughters. "He didn't want Bassons to win, just didn't like him, I guess. I was against us pursuing because we didn't have the yellow jersey and it wasn't our responsibility. Also it was a flat stage and by catching Bassons, we would just set things up for the sprinters and we didn't have a sprinter. There was also a bit of self-interest. I was pretty low physically and a long pursuit wasn't what I needed." Armstrong relented and Bassons got his win. Finally, those who hoped cycling was changing had something to cheer. In the French sports daily *L'Equipe,* the newspaper's chief cycling writer, Philippe Bouvet, didn't hide his enthusiasm: "Isn't Bassons's victory one of the most important of the entire season?" he asked rhetorically.

It was a false dawn. With the best teams in the world ready to compete, the 1999 Tour de France played to a new, faster rhythm. Bassons's version of that story was told in the diary he wrote for the daily newspaper *Le Parisien* and in the many interviews he did with journalists. He had begun the season with optimism, but now, six months later, he was much less hopeful. France's longitudinal testing highlighted a number of irregularities but no one was being punished. Bassons also reminded his readers what it felt like to be part of the peloton. "We are racing at an average speed of more than fifty kilometers per hour, as if the roads of France are nothing more than one gigantic descent."

Le Parisien presented Bassons's column with the introductory sentence saying: "Bassons rides the Tour on pure water; that is to say, without doping products." For many of his peers that sounded like a taunt, and during the first few days fellow riders called him "the chronicler" and "the clean rider" without hiding their derision. There was the occasional exception. On the stage from Laval to Blois, the German rider Udo Bolts approached Bassons. "Your position on doping is a good thing," Bolts said. "It shows you have character." But as Bassons continued to offer frank views, he alienated more of his peers. He also realized that once the race begins, the Tour de France family circles wagons and unites in their defiance. When it speaks, the Tour prefers to speak with one voice. By continuing to pen opinions that hurt the family's interests, Bassons became more and more isolated. Ridicule turned to open hostility.

One day Pascal Chanteur, a French rider with the Bonjour team, spoke to him. "You've got to stop your bullshit. You're on your own; you've turned everybody against you. What you're doing is wrong. Journalists are idiots." Meaning well, the veteran rider Thierry Bourguignon, who had been a teammate at Bassons's first pro team, Force Sud, tried to offer good advice. "The journalists are just using you," he said.

"I know that but I am also using them to say what I have to say."

"Why do you always mention me in your column?"

"Because you are the only one who continues to speak with me."

The irony was that as Bassons became a pariah among his colleagues, the public warmed to his fearlessness. Before the race he was barely known, but through the first ten days, he was applauded wherever he appeared in public. Unsure about its heroes, the public embraced the antihero. But their support could sustain him only up to a point because Bassons's disillusionment deepened as the race progressed. Nine days into the race he rode the individual time trial at Metz in the morning and watched from his hotel bedroom as the Tour favorites rode theirs in the afternoon. They raced at a speed he barely recognized and did so with apparent ease. Disgusted, he zapped the remote control and watched Formula 1's British Grand Prix at Silverstone, and then, driven by some morose desire to torture himself, he returned to the Tour to see again just how fast the favorites went. He took a particular interest in Armstrong's performance and not solely because he produced the fastest time of the day.

When he looked at Lance Armstrong, Bassons saw a rider physiologically similar to himself: same height, same weight, and if Bassons's VO_2 Max was superior to Armstrong's (eighty-five to eighty-three), their capacities did not seem that different. Yet, in terms of performance, they came from different planets. Two days after the Metz time trial, Armstrong crucified his rivals on the key mountain stage to Sestriere in the Italian Alps. Bassons and his teammates watched television highlights that night and couldn't believe the ease with which Armstrong rode away from his rivals on the last part of the climb. They talked about quitting the Tour in protest and none of those who watched believed in the authenticity of the performance. "Did you see how he reeled in those guys without even opening his mouth? That's impossible," one said. "And how he dropped the others without forcing?" said another. "Can you believe it?" Bassons's column in *Le Parisien* reflected his continuing disbelief.

The next day, July 14, was France's national holiday commemorating the storming of the Bastille in 1789. Bassons apart, the spirit of rebellion was nowhere to be found. Whatever was said behind closed doors the previous evening, the Tour again spoke with one voice the following morning. The race was at the halfway point, Armstrong was in control, and, within the family, it was time to unite around the champion-elect. In the way of professional sport, there was also a belief that Armstrong would be the charismatic leader cycling needed in the wake of the 1998 scandal, a star well capable of attracting new sponsors to the sport. Those riders who doubted the worth of his ride to Sestriere the previous evening woke up to a new day with a new perspective. It just made Bassons madder.

As the new patron was entitled to do, Armstrong decreed the next morning that the pace on the journey from Sestriere to Alpe d'Huez would be sedate until the approach to the first climb. Soon after the start, Bassons defied Armstrong by attacking and was immediately pursued by the U.S. Postal Service team. When they joined him, Armstrong put his hand on Bassons's shoulder and spoke to him in English.

"What are you doing?" he asked.

Bassons knew enough English to reply. "I make the race, I attack."

"You know, what you're saying to the journalists, it's not good for cycling."

"I am simply saying what I think. I have said there is still doping."

"If that's what you're here for, it would be better if you returned home and found some other kind of work."

"I am not going to leave when I haven't changed anything. If I have things to say, I will say them."

"Ah, fuck you."

Bassons was then reproached by his teammate, Stéphane Heulot, who said he should stop causing trouble. Although Heulot had been one of those ready to pack his bags and leave the Tour in protest the previous evening, his resentment was short-lived. Coincidentally, when Armstrong gave the go-ahead for the race to begin in earnest, it was Heulot and his compatriot Bourguignon who composed the two-man breakaway that animated the day's race, and their adventure endured for more than ninety miles. Their Bastille Day daring cheered the home nation and lasted until the big boys decided enough was enough. They were reeled in on the higher slopes of Alpe d'Huez. Heulot hung on for eleventh place—one minute and forty-five seconds behind the day's winner, the Italian Giuseppe Guerini.

By now Bassons was an outcast, despised by the pack—and if there were any dissenters, they weren't foolish enough to identify themselves. For the first time in the race, his morale was low. At the team hotel in Saint-Etienne thirteen days into the race, Marc Madiot decided to confront what he felt had become a serious problem for the team. Because he was the enemy, it was impossible for Bassons to achieve anything in the race. If he tried to get into a breakaway, he was chased down, and in any case, no breakaway can be successful without help from friends in the peloton. No one was going to work with *Monsieur Propre.* And for Madiot, that was the lesser part of the problem, because it was clear the team was also suffering from the rebel's outspokenness. He talked with Bassons during his evening massage and explained that, so far in the Tour, he hadn't stopped him from speaking about doping, but now he was going to be firm. He did not want Bassons to talk about doping, and if the journalists asked, he was to say that he was at the Tour to race and that was it.

"I don't agree," said Bassons. "Why should I be silenced like the others? From the start, I have not pointed the finger at anyone. I have been asked questions and I have said what I think, that's all. For the last two years I have said nothing about this problem, obeyed the law of silence, but doping has really hurt my career and now I am going to say what I think. I am not going to be stopped."

"Don't you see," replied Madiot, "you are turning everyone against you. I said to you before the start that if you waited until after the Alps, you would be able to get in a breakaway, but if you continue to speak, no one is going to let you. You are in the process of torpedoing yourself and your team." They argued a little more and then Madiot's voice mellowed. "Look, I accept what you're saying is true, but there are enough problems, enough bad publicity for the sport. We shouldn't make things worse." Bassons replied he wanted to finish the race with a clear conscience and that was more important than winning a stage. They argued again later in the evening when Madiot, in front of the other riders, told Bassons he was the reason Heulot had not won at Alpe d'Huez. This was a far-fetched claim since Heulot, in a state of exhaustion, had lost one minute and forty-five seconds on the leaders through the final three miles of the climb. Asked by Bassons if he believed Madiot's analysis of the Alpe d'Huez stage, Heulot replied, "I don't know. It's possible."

Once the pressure began to come from within his own team, Bassons cracked. He was the first to leave the dinner table that evening, and by then, he knew he wouldn't start the next day's stage. Telephone calls to Pascale, his girlfriend, and to Antoine Vayer, his trainer, resolved nothing. Vayer wanted him to reflect more and to leave the decision until the morning. "After a good night's sleep?" asked Bassons, incredulously. He had phoned from a booth in the hotel lobby, and when he went upstairs, his roommate, Heulot, was watching television. Not a word passed between them. The next morning Christophe Bassons abandoned the Tour.

On the road to Saint-Flour, the stage-end town to which he should have been racing that day, a banner was draped across the road: FOR A CLEAN TOUR, YOU MUST HAVE BASSONS.

Chapter 12

FRANKIE'S BREAKING POINT

It takes a lot to crack Frankie Andreu.

For twelve years, he earned his living as a professional cyclist, and for most of the last twenty years, he has worked with people who shared his passion for the sport. Using the fingers of one hand, you could count the times when he wondered why he'd gotten involved, and you would have fingers to spare. Perhaps this was because he was hooked on the buzz from riding his bike; feeling the wind on his face, the sun on his back, knowing that five hours on a bike stilled the mind's restlessness. He hasn't found much in life to compare with that. And he got on well with the people drawn to cycling; mostly they're his kind of people—kindred spirits, you could say—and through the years, he never tired of being among them. About the truest thing you could say of Andreu also happens to be a generous compliment: he never had a teammate who didn't like him. Or, if there was one, he hasn't owned up. They liked Frankie because he was a regular guy, without any pretension, and most of all, what you saw was what you got. Teams needed guys like Andreu.

All of which is not to say he wasn't human. It may have taken a lot but Andreu did have a breaking point. There was the time he rode for the French team Cofidis and things got so bad that Frankie completely lost it. It was 1997, the first year of the team's existence, and to put it gently, there were teething problems. Andreu lived in Nice but the team routed every journey through Paris. Say he wanted to fly to Barcelona. He couldn't just go Nice-Barcelona; it had to be Nice-Paris-Barcelona, and by the way, it had to be with Air France. When the team went to the Belgian classics, it didn't actually go to Belgium. Instead it set up camp on the French side of the border and endured the one-and-a-half-hour car journey to the start of the race so they could sleep under a French sky and could eat French croissants the next morning.

Part of the reason Andreu signed for Cofidis was to continue riding with Lance Armstrong, but testicular cancer prevented his friend and former Motorola teammate from riding for the French team, and that left the Swiss rider Tony Rominger and the Italian Maurizio Fondriest as the new leaders. Rominger was getting toward the end of his career and Fondriest was injured for much of the season. When Fondriest returned to action, his teammates wished he had been injured for all of the season. Whereas Armstrong was a natural-born leader, Fondriest was a different animal; he commanded a big salary but no respect. Cofidis worsened a bad situation by agreeing to use bikes manufactured by Fondriest's cycle company, and it was the bikes that found the fault line in Andreu's seemingly implacable nature.

He thought they were terrible. These were first edition Fondriest bikes, so flexible the riders would derail the chain by pushing big gears, and to Andreu the bike felt as secure as a piece of cooked spaghetti. The complaints were endless but nothing was done. Fondriest said the bikes were fine, and that was it—end of discussion. It drove Andreu crazy and in the end he flipped. Midway through the season, he and Kevin Livingston were packing their bikes in Nice, getting ready to depart for a race. Andreu looked at the frame, felt again his hatred, and it set him off, talking about how he didn't want to race on it, how he would do anything to get a new bike, how the team just had to do something. And then, he cracked.

Frankie was one of the guys who kept his bike well, and even when handing it over to the mechanics before a race, he gave it to them spotlessly clean. Now it was tidily packed in the bike bag, ready for the journey to the airport, but Andreu was about to treat it as never before. Taking the bike from the bag, he walked from his apartment to the nearby sidewalk. Livingston watched in a state of temporary bewilderment, knowing something was about to happen but momentarily immobilized. Reaching the sidewalk, Andreu caught the frame by the front forks, lifted it high over his head, and slammed it onto the cement curb. Realizing he had badly buckled the chain stays, he uttered a quiet "Perfect." Livingston could barely believe what he had witnessed, and, fearful of the consequences, he babbled on about the impossibility of explaining this to the team bosses—until he was interrupted by Andreu's cool matter-of-factness. "The airline," he said. "I will just say the airline damaged it." So simple, so plausible—they both laughed, the crazy laughter of the irrational. Eventually they put the damaged frame back into the bag and were about to depart when Livingston asked Andreu how he was going to explain the dirt and dust of the cement curb. "Good point," said Andreu, who took the frame from the bag, cleaned off the dust, and then set off. In Spain, where they were due to race, Andreu and Livingston handed over their bikes to the mechanics. An hour or so later, one of the mechanics came to Andreu's hotel bedroom: the airline had damaged the frame. "No, really?" Andreu said. The mechanics were sorry but it wouldn't be possible to get a replacement Fondriest bike in time for the race.

Would he mind riding a different frame?

The second time Frankie Andreu cracked, it was a more serious matter.

It happened in the aftermath of the Festina doping scandal of July 1998. Some riders didn't think that was much of a scandal at all. Festina had got caught with drugs. Their problem. Others saw it as the shuddering collision with reality needed to make the sport recognize its pervasive doping culture. Andreu was in that camp. Through the midnineties, he saw the way racing changed and heard all the talk about r-EPO. He tried not to believe the worst of it and said it was hard to separate rumor from fact. Riders were always convinced the guy on the other team was doing it but no one was coming forward with his hand up. Yet he couldn't delude himself—r-EPO worked, it was undetectable, and the UCI's decision to impose short-term suspensions on those with hematocrits exceeding fifty made it seem all right to use the drug up to the fifty threshold. He thought of it as legalized doping.

But he wasn't going to get too down about it. Through the days of rampant abuse in the midnineties he didn't do r-EPO, but that was okay, he wasn't paid to win and he could do his équipier's job without the drug. Instead of worrying about others, he concentrated on what he could control. Prepare properly, ride hard, give everything, and accept the results that follow. In a curious way, it was easier to do this riding for the U.S. Postal Service team. Armstrong expected team riders to be just that, team riders. There was one leader, him, and for anyone else to think of personal glory was an affront to the ethos of the team. In return, Armstrong won when it mattered and his équipiers made good money. From Andreu's point of view, it was a fair exchange.

But when the Festina guys got found out, he thought things would change. Surely no sport could just carry on in the old way after police searches revealed widespread doping. It wasn't just Festina; Spain's number-one team, ONCE, was accused of dealing in drugs; suspicions existed about many other teams; and one rider, Rodolfo Massi, was arrested on suspicion of supplying drugs to his fellow riders on the French-based Casino team. Massi would eventually be acquitted of all charges but the suspicions and arrests were destroying cycling's credibility. "The sport," said Tour de France chief Jean-Marie

Leblanc, "needs a new morality." Through the winter of 1998, that became Frankie Andreu's hope.

Believing change was necessary wasn't enough to bring it about. There was still no test to detect r-EPO or human growth hormone, and if used in a measured way, other drugs were difficult to find. What Andreu realized in the early races of the 1999 season was that nothing had, in fact, changed. Well, one thing had changed: even among themselves, riders became more cautious when speaking about doping. Willy Voet and Festina, the villains of the previous summer, were not spoken about, and the legacy of the Festina scandal was greater secrecy. The law of silence was reinforced, and lacking the will to learn from the mistakes of the past, the sport went on repeating them. The evidence of continued doping was everywhere. Sprinters flew up first category climbs, new young professionals adapted to the rigors of their first year in the peloton with absurd ease, and Frankie Andreu fell farther and farther behind. It wasn't so bad in one-day races—he could get by in those—but it was much tougher in stage races. Standing at the start line on the third morning, with two days' tiredness in his legs, he looked around and saw rivals still fresh and, he imagined, fueled by abundant red cells. This wasn't something he could prove but he just knew it. And in the spring of that year, 1999, it got to him.

"I was tired of having no chance, tired of getting dropped, tired of seeing others take advantage of me," he said. "I just wanted to feel normal again, to get back my place in the peloton. This was something I had worked hard to achieve and through no fault of mine, I had lost it. I was training as hard as ever, I was feeling as well as any other year, and I was racing great. The problem was I just couldn't keep up. I wanted to ride to my abilities, to be the team rider I had been throughout my career, but the difference was that I now needed r-EPO to accomplish that."

Andreu had heard enough of the stories to know Switzerland was the place to buy r-EPO, and as he lived in Nice, it wasn't that long a journey. He was told you didn't need a prescription, not even a doctor's note, and, probably most important of all, no one was going to ask to see identification. All you needed was money. But it took weeks for him to finally cross the line. Back and forth, he wrestled with the decision. "Should I, shouldn't I? One day it was 'yes,' the next day 'no.' " He told himself he was doing it to keep his job with the U.S. Postal Service team, to go on providing for his family, but he knew Betsy, his wife, wouldn't buy that. She would say it was cheating and therefore immoral. It wasn't something he could discuss with her before doing it.

In the end, the Tour de France made up his mind. Never before had he ridden in a team that had a potential winner. You might have made a case for Andy Hampsten in 1993, but he had been a long shot against the Spaniard Miguel Indurain. For this Tour, Armstrong was one of the favorites. But to win, he would need the support of a strong team and that meant every rider. The responsibility preyed on Andreu's mind. This was the race where he had to do his job, where he had to be strong. Once the decision was made, it seemed like a weight lifted from his mind and he could get on with preparing for the race.

He heard how simple it was to drive to Switzerland, walk into a pharmacy, and just ask for r-EPO. Little things revealed his reservations: he didn't ask the Swiss pharmacist for r-EPO but for one of the brand names under which the drug was sold. That felt easier. Immediately understanding what Andreu wanted, the pharmacist asked, "How many boxes?" "Just one," Andreu replied, noticing the pharmacist's look of exaggerated surprise. "You don't need more than one?" he asked. Andreu picked up on the inference and said firmly, "No, one is just fine." The journey back into France was one he would prefer not to repeat.

All the way to the Swiss-French border, his heart raced as he considered the consequences of being stopped and searched. Should he immediately admit to having the r-EPO, or could you try to act innocent and hope not to be searched? In the end, the border guards just waved him through and he was relieved to arrive back at the apartment in Nice. Though he had heard a lot of talk about r-EPO, he didn't actually know how the drug was used. How many injections each week? How much in each dose? How long would it last? At what temperature should it be stored? Eventually, he got some advice and began to take the drug leading up to the 1999 Tour de France. What intrigued him was that a tiny amount of a clear liquid, injected into the shoulder, increased his red blood cells. He used the drug at the apartment in Nice, refusing to take it to the preparatory races and the Tour. That would have been risky. He didn't tell people what he was doing and no one asked him. It was his secret.

At first there was no discernible effect on his performance, but once the Tour de France got going, he began to notice the change. Slowly, he felt stronger and better able to do the équipier's job that had once come easy to him. The drug returned him to his rightful position in the peloton, that of the respected team rider. There was no sense of guilt because most riders were using r-EPO, and he also knew there were plenty of guys doing far more. Many were using human growth hormone, corticoids, testosterone, and other products, but for him, that would have overstepped the mark. All he wanted was something to allow him to ride at his normal level.

He didn't suddenly start winning stages or leading the pack in long pursuits of breakaways. In fact, apart from himself, perhaps no one saw any difference in his performance. But there were differences. Instead of finishing each day in a state of total exhaustion, he was just ordinarily tired and he woke the next morning feeling his body had recovered a little. Andreu didn't imagine he would spend more than two weeks defending the race leader, but that's how it panned out in '99 and he needed his strength. He still suffered in the mountains and there were times when he struggled on flat roads, but because his body was recovering, he was able to do his job. That pleased him. Pretty much from the moment he had turned professional eleven years before, he understood the level of his talent; it wasn't enough to make him a star or win more than a handful of races, but he was strong and got the credit due a loyal team rider. When r-EPO came, it took that away. His weakness was to want that back, to be able to do his job and make sure no one could say he wasn't pulling his weight. Respect, he supposes, was what he was after.

After her experience in the hospital room in Indiana in late October 1996, Betsy Andreu's view of professional cycling changed. It was like her eyes had been opened, and thereafter, she couldn't close them. As she told her then fiancé, she wasn't prepared to marry or be married to a doper. Frankie convinced her he didn't do performance-enhancing drugs and she felt he was almost as shocked as she had been when Armstrong made his admission in Indiana. She later asked her husband if Armstrong had subsequently spoken to him about the hospital room consultation. Andreu told her that he had. He said, "Lance wanted to know how you reacted. I told him, 'Not good.' "

In 1998 Betsy Andreu traveled with Kevin Livingston's wife, Becky, to the Tour de France at a time when riders were staging a sit-down strike at what they saw as the heavy-handed tactics of French police. She believed that if riders were that upset by the police searches, they must have had a lot to hide. She also believed clean riders would have welcomed police involvement. For the 1999 Tour, Betsy Andreu's plan was to travel from her home in Dearborn, Michigan, to France for the final week of the race and to be on the Champs-Elysées on the final Sunday afternoon. Two months before the start of the Tour, little Frankie Andreu was born and Dad got to spend just a little time with him before he had to leave for France. Almost seven weeks passed before father and son were reunited.

Before departing for Europe, Betsy Andreu watched television coverage of the first two weeks of the race. "One stage in particular is embedded in my mind," she said. "It was the first mountain stage, the one to Sestriere, and as they began the climb, Frankie was at the front of a big line of U.S. Postal Service riders, setting the tempo. Frankie is a domestique and about as much a climber as the Pope is an atheist. 'What the hell is this about?' I said to myself as I watched.

I called up a friend of mine in Paris, Becky Rast, and asked if she was watching the Tour. 'Yeah,' she said, 'isn't Frankie doing great?' I yelled back, 'Frankie isn't a climber. What the hell is he doing pulling at the front on a first category mountain?' "

If you are the wife of a domestique, or team rider, you don't often get to see your man on television. For about a minute or so, at the beginning of the climb to Sestriere, Frankie Andreu dominated TV coverage of the race, and his wife, watching on the other side of the Atlantic, felt nothing but dread. What she saw convinced her Frankie had caved in to cycling's doping culture. After he got back to the team hotel, Betsy called her husband. "So, how did you manage to be leading the peloton on a first category climb?" she asked, making no attempt to disguise her skepticism. "I fell back right after that one effort," he said, but she didn't feel he should have been there in the first place. "Sure, Frankie," she said, and the conversation soon ended because in his state of mental and physical exhaustion, it wasn't something he felt like discussing.

He could have said he was on the front for about a minute and that immediately after he dropped back, George Hincapie took over, and even though Hincapie wasn't a climber, he led the pack for a far longer stretch. He could have pointed to Armstrong, Livingston, and Hamilton, who had been to the Alps with Michele Ferrari in the weeks before the race, and pointed out how strong they had been from the start. But Andreu would not point fingers, not to his wife, not to anyone. In any case, it wouldn't make the slightest difference. His wife didn't care what the others did, that was their business. *He* was her husband.

Betsy Andreu, her mom, Betty Kramar, and two-month-old Frankie joined up with Frankie senior in Carcassonne close to the foothills of the Pyrenees, a week from the end. Betsy's reservations about what had happened on the ascent to Sestriere were not discussed, as a weary dad marveled at how much their son had grown in seven weeks. They reunited on the Champs-Elysées the following Sunday, an hour or so after Armstrong and the U.S. Postal Service team had ridden triumphantly into Paris. Frankie Andreu was elated. After six Tours in which he was an unknown équipier, he was now a member of the team that delivered the new champion, Lance Armstrong. It felt good. Speeding down the Rue de Rivoli, wheeling left onto the Place de la Concorde before making the right turn onto the Champs-Elysées and up to the finish line, he felt a tingle of excitement he hadn't ever experienced in a bike race. He'd ridden hard and given everything. With his Postal teammates, he led the pack onto the Champs-Elysées, the yellow-jerseyed Armstrong in tow, and that was good. As good as it gets. At that moment, r-EPO just didn't come into it.

That night the Postal team held its post-Tour dinner at the Musée d'Orsay, the world-famous art museum that was once a railway station. An estimated three hundred guests attended and it was a far cry from the Mexican restaurant used during the Motorola years. During the evening, Frankie asked Betsy if she had congratulated Lance on his victory. She replied she hadn't. He asked if she would. Betsy refused, saying she didn't believe Lance won the race clean. They argued and Frankie was adamant she should offer her congratulations. But she didn't believe in the Tour, didn't believe what she saw her husband do on the climb to Sestriere, and didn't believe in the winner. If her husband had succumbed to the pressure to use some banned drug during the Tour, it was only so he could help the team leader. To her, it just wasn't plausible that good team riders would need to dope to help a clean leader. So she didn't congratulate Armstrong.

The next day the Andreus traveled south to Nice, where Frankie would recover from his Tour exertions. Betsy loved that time; Frankie's training rides were relatively short, all three of them were able to spend some time on the beach, and most of all, Betsy was able to go to the supermarket alone while Frankie took care of his son. After one such outing, she was unpacking the groceries when she noticed the small thermos in the refrigerator. "What's that?" she asked Frankie, who was tending little Frankie. "Oh, nothing," he said. But before he could do anything, she opened the thermos and saw a vial of r-EPO left over from the box bought in Switzerland. Seeing the clear liquid in a vial, she became upset and angry. Afterward she would hate the way she scolded him, detest the way she bitched and nagged and told him how wrong he was to have used that shit. She would regret her lack of appreciation for the effort he had made in the Tour de France.

"What am I supposed to do?" he said. "Everyone's on this shit and more. I did the bare minimum. I wouldn't go to see Ferrari, I wouldn't get on a program, I don't want this shit in my body, but if I couldn't keep up in the Tour, I wouldn't have a job. What do you want me to do?"

"Get off the U.S. Postal Service team. Just get off it."

"Hon, this problem is on every team," he said.

"You've got to promise me you won't do it again. I don't want you to do this, even if you don't ride as well."

"I am not going to do it again," he said.

It would be wrong to say Frankie Andreu regretted taking r-EPO for the 1999 Tour de France, because at the time, that wasn't how he felt. What he wanted was to be himself, to ride at his level, and the sick thing about the sport at that point was that you couldn't ride at your normal level without r-EPO. Was it wrong to want to be the rider he had always been? The irony was that he never used the stuff again. He didn't think it was healthy, didn't like the deceit, and he wasn't prepared to lie to his wife. She saw it like it was black and white. He knew there were many shades of gray, but one of the things he loved in her were her moral values. With Betsy, it was right or it was wrong.

And, deep down, Frankie knew it was wrong.

Chapter 13

A STRANGE KIND OF GLORY

Lance Armstrong's win in the 1999 Tour de France race was acclaimed not solely because it was a formidable athletic achievement, but it was also a triumph of the human spirit. The winner wasn't simply a talented sportsman at the peak of his powers but a twenty-seven-year-old who, three years before, had been left at death's door. In the league of near-miraculous recoveries by sportsmen, Ben Hogan's return from a horrific automobile accident in 1949 and Greg LeMond's comeback from a shooting accident in 1987 were right at the top. Hogan won six golf majors after making a good if incomplete recovery from his injuries. And when all but the last thirty-eight pellets were taken from LeMond's organs, he recovered well enough to go back to Europe and win the Tour de France for a second time, and then a third. But Armstrong went further. His midcareer duel was with life-threatening cancer. To win that battle and return to elite competition was extraordinary, but to then conquer the Tour de France made his the greatest comeback of all.

His victory transcended bicycling, even sport itself. It was a story of hope for those with no interest in the Tour de France. Not everyone understands what it takes to win the Tour, but there is universal understanding of cancer's deadly powers and eternal admiration for those who survive the disease. Armstrong's triumph proved not only that there was life after cancer but that it could be greater than the life that preceded the illness. His story crossed all frontiers and touched millions. The brash kid who turned up in Europe in the early nineties with his own business card and branded baseball caps was ready for the avalanche of opportunities. TV ads, multimillion-dollar sponsors, endless television coverage, a bestselling autobiography, and—overnight, it seemed—Lance Armstrong became an icon. Bill Stapleton, his manager and friend, articulated the impact of the '99 Tour victory in an interview with Mike Hall, writing for the magazine *Texas Monthly* in June 2001. "In the beginning we had this brand of brash Texan, interesting European sport, a phenomenon," said Stapleton. "Then you layered in cancer survivor, which broadened and deepened the brand. But even in 1998 there was very little corporate interest in Lance. And then he won the Tour de France in 1999 and the brand was complete. You layered in family man, hero, comeback of the century, all these things. And then everybody wanted him."

Yet it would be wrong to say the acclamation was unqualified. The drug scandal of 1998 tempered some reactions to Armstrong's victory in 1999 and the closer one got to the heartland of European cycling, the more one found pockets of skepticism. Tour de France chief Jean-Marie Leblanc wanted the 1999 race to be a fresh start and called it the Tour of Renewal, hoping the illumination of cycling's dark secret in 1998 would lead to moral regeneration. Before the '99 race reached halfway, Leblanc lowered his sights and talked of a Tour of Transition. He knew the doping problem had not gone away, but by the end, he said he believed Armstrong was a worthy champion. Others weren't so sure.

Two of the more serious French newspapers, *Le Monde* and *Libération,* came to the same conclusion as Frankie Andreu and Christophe Bassons: in cycling nothing had changed. Writing in *Libération,* the philosopher and author Robert Redeker lamented the Tour no longer tugged at childhood dreams. "We now think of the Tour de France as a paradise lost, like the tales of a grandmother around the fireplace or at the foot of a bed. . . . A huge gulf exists between the race and the racers, who have mutated into PlayStation characters, while the public, the ones at the folding tables and the tents, drinking pastis and fresh *rose du pays,* are still real. This chasm, harbinger of the death of the Tour as we have known it, highlights the difference between the athletic types promoted by today's race and the people's men of the past, born of toil, hardened to pain, adept at surpassing themselves (Gino Bartali, Jean Robic, Fausto Coppi, Louison Bobet) who have been replaced by Robocop on wheels, pharmaceutically manufactured men, to whom no fan can relate."

For evidence of change, one looked at the average speed of the 1999 Tour. Jean-Marie Leblanc said he would not be displeased if there was a slowing down, reasoning that with r-EPO less in use, the speed had to fall. The '99 Tour ended up being the fastest in history. Others looked at Armstrong's record in the Tour de France before testicular cancer, where his only finish from four starts was a modest thirty-sixth, and found it hard to believe the same rider could be so dominant in '99. Armstrong sensed the skepticism and offered a vigorous defense. He said he was "one of the most tested athletes in the world" and he reminded people he had been a winner from the moment he began competing in the sport. The difficulty was that passing drug tests was not proof of innocence, as the Festina riders proved. They were part of a systemic doping program without anyone failing a drug test.

Those concerned about doping looked for other clues to the ethics of a team. In this respect, U.S. teams held an advantage precisely because of where they were coming from: America did not have a history of doping in the sport. 7-Eleven and Motorola, the first U.S. teams to campaign extensively in Europe, retained a distinctly American ethos by recruiting U.S. and other English-speaking riders. Mike Neel and Jim Ochowicz, both Americans, were the respective directeurs sportifs, and the team doctor was the English-speaking Americanized Italian, Max Testa. But the U.S. Postal Service team traveled a different road from their predecessors. They employed a European directeur sportif, a European doctor, and European soigneurs. Johan Bruyneel, Postal's Belgian-born directeur sportif, spent much of his career with the Spanish ONCE team, who had been investigated on suspicion of doping and whose longtime directeur sportif, Manolo Saiz, would be one of those arrested in the Spanish doping investigation, code-named Operation Puerto.

For those wishing to understand Bruyneel's attitude toward doping, there was a revealing interview with Samuel Abt in the *International Herald Tribune* four weeks before the start of the 1999 Tour. At one point in the interview, Abt referred to cycling's image as a doped-up sport.

Abt: This is not a good time for the sport, is it?

Bruyneel: For me, the situation is very simple: Cycling is a sport in a very bad light, and the reason we got there is the fact that, three years ago, the riders accepted too easily the fact that the authorities could install blood controls. Having these controls would have been a very good thing if it had not been done only in cycling. Now there are more and more controls and the image of cycling is worse and worse. And I have no reason to believe that other sports are cleaner than cycling. Tell me one sport where the drug test is better.

Abt: But the riders, led by the Italians, proposed the blood controls themselves, didn't they?

Bruyneel: Yes, but who were those riders? They were riders near the end of their careers.

Further on in the interview, Bruyneel questioned the motives of those investigating doping in cycling. "One problem has been that people in the police and justice system have seen that by opening a case, they can have a lot of publicity for themselves." Abt pointed out that the prosecuting magistrate in the Festina affair, Patrick Keil, refused to grant any interviews and had not allowed himself to be photographed. "Yes," replied Bruyneel. "He's a well-known person now. Before he was nobody." The blood tests that the Postal directeur sportif believes should have been resisted showed cyclists had elevated hematocrit levels and dangerously high iron stores, both indicative of widespread r-EPO abuse. The investigations were opened because police in France and Italy found illegal performance-enhancing drugs wherever they looked.

At first Armstrong reacted with hostility to the questions and the skepticism. *How dare they?* When *Le Monde* investigated his positive test for cortisone at the 1999 Tour, he said the newspaper was "the gutter press." Perhaps in the United States, where one French newspaper might not appear that different from another, he could get away with that dismissal, but in Europe, the accusation was laughable. *Le Monde* is arguably the most highbrow and high-minded newspaper on the continent. His bullying of Christophe Bassons on the Alpe d'Huez stage of the 1999 Tour also raised questions, not the least of which was this: if Armstrong was anti-doping, why did he feel such antipathy toward a young cyclist who was clearly and passionately crusading for his sport to be drug-free, a young rider who was prepared to sacrifice his career rather than surrender to the sport's drug culture?

In 2000, Armstrong won his second Tour, but rather than go away, the doubters grew more numerous. This was partially in response to a judicial inquiry into the U.S. Postal Service team, opened by the French police in November 2000, following the dumping of medical waste by two Postal team members during the Tour de France four months before. Hugues Huet, a news journalist with France 3 television, was one of those who felt doping was still rife in cycling, and midway through the 2000 Tour, he decided to stake out the hotels used by the Postal team. With a cameraman and a soundman, he set up his covert operation close to the hotels used by Postal. On the first morning they watched as two team employees walked to a Volkswagen Passat TDI in the hotel parking lot, each taking a plastic trash bag out of a backpack and throwing it into the trunk of the car. They recognized the men as Postal team doctor Luis Del Moral and the team's chiropractor, Jeff Spencer. After the Postal men drove away, the journalists followed, tailing them until the autoroute but losing them shortly afterward.

Three days later, on the morning of the Alpine stage from Courchevel to Morzine, Del Moral and Spencer set off in their Passat on a route to Morzine that avoided the traffic congestion caused by the Tour cavalcade. This time Huet used three cars in the tailing operation, and after driving for more than an hour and a half, they filmed Del Moral and Spencer dumping their trash bags at a roadside rest area near the town of Sallanches. After the Postal staffers moved on, Huet and his team recovered the trash bags and examined their contents. Mostly what they found were legal recovery products, but they also found 160 used syringes, many blood-soaked compresses, used IV equipment, and a questionable product, Actovegin. "One hundred and sixty syringes," said the ex-Festina soigneur Willy Voet, "corresponds to four or five days' usage. Each rider would receive four to six injections daily."

The judicial investigation opened in November 2000 and was finally closed in August 2002. Twenty-one months of inquiry raised many questions but produced no answers, and in the end, the police couldn't produce evidence of wrongdoing and the case was closed. But the team's reputation was damaged and in the months following the opening of the inquiry, Armstrong tried to be more conciliatory in his response to those skeptical about his and the team's success. In April 2001 Bill Stapleton telephoned the author of this book and offered an interview with Armstrong that could deal with whatever questions there were. We met at a hotel in the southwest of France a week later, our first one-on-one interview since the 1993 Tour de France, eight years before. Armstrong asked if Stapleton could sit in on the interview and the attorney did, only once interjecting to clarify a not-very-significant point. The interview dealt almost exclusively with doping.

Q: Cycling is a sport with a doping subculture. When did you start to become aware of that?

A: If you're asking when was the first day that I realized that perhaps this exists in our sport, I don't know the answer because Motorola was white as snow and I was there all the way through '96. Riders like Steve Bauer and Andy Hampsten, these guys were very admirable professional, clean riders.

Q: What about the 1994 Flèche Wallonne race, when the Gewiss riders finished first, second, and third? Not everyone thought that was normal. How did you feel about it?

A: At the time, I was frustrated because I was in the rainbow jersey [denoting world champion] and I was close to making that move. Teams have an ability to ride. They [Gewiss] already had had a phenomenal spring, so things had been growing, growing, growing. Once a team starts winning—you see it every year, there's always a team that comes out and they get a couple of wins early on and then they get a couple of big wins and the next thing you know, they're on top of the world, and I believe in that momentum.

Q: Are you saying on the morning after that race, you were 100 percent a believer they had done it clean?

A: The next morning there was obviously articles and people said certain things. If I have to look at that guy and say, "They're cheaters, he's a cheater, the team's a cheater," how could I get up every day and go do my job?

Q: Their doctor, Michele Ferrari, made his famous statement on the evening of that race about r-EPO being no more dangerous than orange juice. Do you remember your reaction to that?

A: (Long pause) Ahmm, no.

Q: You didn't wonder what r-EPO was?

A: I think that sometimes quotes can get taken out of context and I think even at that time I recognized that.

Q: Ferrari didn't come along afterwards and say, "I never mentioned this drug." [The doctor actually said he had been quoted accurately in *L'Equipe.*] Had you been aware of r-EPO?

A: Here we're talking seven years ago. Had I heard of it? Probably.

Q: Ferrari was, in effect, saying he gave it to his guys?

A: I didn't read the article, I don't know.

Q: It is obvious to everybody r-EPO use became a big thing in cycling in '95 and '96. How conscious were you guys in Motorola that r-EPO had become a factor in race results?

A: We didn't think about it. It wasn't an issue for us. It wasn't an option. Jim Ochowicz ran the program that he set out to run, a clean program. Max Testa, the doctor, set out to run a clean program, and that wasn't part of our medical program.

Q: You must have been frustrated at the thought that these guys were using a substance which Ferrari had talked about?

A: There's no proof of that. I wasn't going to sit around and talk about it. This is ages ago for me; that part of my career, that part of my life is finished.

Q: Did you know that Kevin [Livingston, fellow U.S. Postal Service rider] was linked with the [police] investigation into Michele Ferrari in Italy?

A: Yes.

Q: Did you discuss it with him?

A: No.

Q: Never?

A: (Nods his head.)

Q: Even though you would have known he was a rider who was on Ferrari's books, that was printed in loads of newspapers?

A: You keep coming up with these side stories. I can only comment on Lance Armstrong. I don't want to speak for others. I don't meddle in [other] people's business.

Q: A guy who is your best friend?

A: In an indirect way, you are trying to implicate our sport again.

Q: I would have thought it natural you would say, "Kevin, what's this about? Did you go to Ferrari? Is this being made up, did he put your name in his files when you never visited?" You never discussed it, ever?

A: No.

Q: Would it shock you to realize that there are Ferrari files on Kevin that indicate he was using r-EPO?

A: I wouldn't believe that.

Q: Even if you saw the files?

A: I wouldn't believe that.

Q: There are files I have seen where Kevin's hematocrit is listed for July 1998 at 49.7. The previous December it is listed at 41.2. Most medical people say a near nine-point difference in a hematocrit level in a six-month period is highly unusual.

A: I haven't seen the files. I don't know.

Q: Did you ever visit Michele Ferrari?

A: I did know Michele Ferrari.

Q: How did you get to know him?

A: In cycling when you go to races, you see people. There's trainers, doctors, I know every team's doctor. It's a small community.

Q: Did you ever visit him?

A: Have I been tested by him, gone and been there and consulted on certain things? Perhaps.

Q: You did?

A: (Nods in affirmative.)

Q: He's going to be tried for criminal conspiracy.

A: I think the prosecutors and judges should pursue everybody, regardless of who it is. It is their job to do that.

At another point in the interview, the involvement of Armstrong's longtime trainer and friend Chris Carmichael in Greg Strock's case against USA Cycling and his former coach, René Wenzel, was discussed. Though named in Strock's original complaint, Carmichael was excluded from the amended version of his statement. In that first document, Strock described a scene at a race in Spokane, Washington, in which he was taken by Wenzel to Carmichael's bedroom and injected with what he says was called "extract of cortisone." The amended court document described the same incident but without identifying "the other coach" as Carmichael.

Q: If this coach is Chris Carmichael, it obviously has a relevance for you?

A: Absolutely.

Q: Because you worked with Chris all the time?

A: Sure.

Q: Have you spoken to Chris about this?

A: I talked to Chris. I mean, Chris is still my main adviser, I talk to him all the time.

Q: Has Chris said to you he definitely wasn't that coach?

A: What's interesting to me is that you're sort of building this up and dramatizing this thing as if he came in and injected the kid with r-EPO.

Q: No, extract of cortisone is what Greg Strock believed it was.

A: There's a very big difference between what he believed it was and what it was. What if it's a B-shot [vitamin B]?

Q: What has Chris said? Have you spoken to him about it?

A: Oh, Chris is absolutely innocent. Chris Carmichael is way too smart and way too aware to give a junior cortisone. I believe that 100 percent. I will believe that till the day I die.

Q: If Chris gave Greg Strock at age seventeen any kind of injection, he was breaking the law.

A: Even a B-shot?

Q: Yes.

A: I'm not aware of that.

Q: It's a question of a man without medical training giving injections to a minor.

A: But what age was Greg Strock in 1991? [The incident actually took place in 1990.] He was eighteen. Therefore he was an adult. Let's get the facts straight. I'm not stupid.

Q: Why do you think Chris wasn't mentioned in the [amended] formal complaint, given the other coach was?

A: You will have to ask Chris or Greg Strock. I'm not their adviser. I'm not anybody's attorney. I don't know.

Q: It's my information Chris has made a settlement with Greg Strock and his legal team. That's why his name wasn't mentioned.

A: Ask Greg or Chris.

Q: Chris didn't explain it to you?

A: As far as I am concerned, it was a case between Greg Strock and René Wenzel.

Q: But this other coach is mentioned and it has been written in an American newspaper that this coach was Chris Carmichael.

A: I have never seen that.

Q: It was written in *The Orange County Register.*

A: I have never seen that.

Q: You are not aware of Chris making any [financial] settlement with Greg Strock or his people?

A: You would have to ask them.

Q: You know nothing about that?

A: No.

Q: If Chris had made a settlement, would you be shocked?

A: Again, I am not his attorney so I don't know the details. Honestly, I don't know the details. I didn't read Strock's deposition. I know he was eighteen at the time because I know how old I was and he was a year younger. Would I be shocked? I haven't even thought about it.

Q: Let's hypothesize. If Chris had paid money to keep his name out of a court case, it would imply he had something to hide.

A: It's a hypothesis.

Q: But it wouldn't look good, would it?

A: At the same time, does it look good that Greg Strock just takes the money? Let's flip it around. Is this about money or is this about principle?

Chapter 14

LEMOND FEELS THE HEAT

Early in 2001, the three-time winner of the Tour de France Greg LeMond heard a story that played on his mind. The information concerned Lance Armstrong, who was then a two-time winner of the French Tour and the natural successor to LeMond as America's greatest cyclist. LeMond wanted to believe in Armstrong's successes in the 1999 and 2000 Tour de France and accepted that his compatriot's improvement since his recovery from life-threatening cancer had been brought about by his twenty-pound weight loss. He understood physiology well enough to know a slimmer and lighter rider stood a better chance on the steep climbs of the Tour de France. But now, doubts niggled LeMond. He heard Armstrong was working with Michele Ferrari, an Italian sports doctor who was then standing trial on doping charges in his native country. Within cycling, Ferrari's name was mud. Why would Lance work with that guy? Around the same time, LeMond was told that at the height of his cancer sickness in October 1996, Armstrong had admitted to doctors at Indiana University Hospital his use of performance-enhancing drugs. LeMond found this hard to believe but what made it stay in his mind was the source, James Startt, an American cycling journalist working out of Paris. Startt was reliable and didn't get much wrong.

In the generation that preceded Armstrong, Tyler Hamilton, and Floyd Landis, Gregory James LeMond from Lakewood, California, was a phenomenon. He had a gift for riding a racing bike that, if lucky, you saw once in your lifetime. Andy Hampsten, a contemporary of LeMond's, said he never saw anyone with anything like LeMond's talent. LeMond's character wasn't dull, either; he was open, vulnerable, optimistic. Long before he won his first Tour de France in 1986, Europeans saw him coming. They considered him *très gentil* and liked that the smile rarely left his blue eyes. They marked him down as the quintessential American child of the sixties—conceived in rock 'n' roll and delivered in a pickup, and they loved him all the more for it. There were parts of him they didn't understand, like the fact that he played golf and went hunting in the middle of the cycling season, but they couldn't see anything they disliked. When he won his first Tour de France in '86, LeMond beat the Breton Bernard Hinault into second place and denied him a record sixth victory in the race. However hard it was for the French to stomach that, they acknowledged the greatness of LeMond's performance. Their love of the Tour was greater than their love for Hinault, and from the moment LeMond had won the Tour de l'Avenir in 1982, they knew his was a special talent. How could they not applaud?

After retiring at the end of 1994, LeMond lost the passion he once had for his sport. Perhaps that is not quite accurate; he did not lose his love for cycling but for the professional branch of the sport. But the doubts had been born before his exit. In 1991 and through '94, he saw change he could hardly believe, speeds that made no sense unless one accepted the existence of a new and devastating form of doping. It wasn't that doping didn't exist in his day, but you could handle guys who used testosterone and corticosteroids. He won his third and final Tour in 1990, and though he was in better shape for the race in 1991, he couldn't compete. "There is one stage I particularly recall," he said. "The average speed was thirty-one miles per hour and that was after we were twice stopped at a railway crossing. Something had transformed the peloton." In the final year of his career LeMond suffered from mitochondrial myopathy, a degenerative muscle disease that drained his strength and may well have been caused by passive doping, because by now the r-EPO era was in full swing and the clean rider was at risk. Cruelly ironic it may have been, but the truth was that if you rode clean against riders using r-EPO, you were so stretched to keep up, you were the one in danger of damaging your health.

Because he was a three-time Tour winner and a double world champion, LeMond's name has long been worth something, and since retirement he has earned his income from his two companies, LeMond Fitness and LeMond Racing Cycles. The latter company is now a division of Trek. There are also the invitations that come the way of an old champion—a sponsor's banquet, a trade show, a guest appearance at a bike race, or some journalist calling and looking for a reaction to a story of the day. LeMond considers himself out of the sport but knows he can never leave it—not absolutely.

His relationship with Armstrong was never close, though they did race against each other in 1993 and 1994 and would occasionally bump into each other in the years that followed. When Armstrong won his first Tour de France in 1999, LeMond says he wanted to believe his countryman won it clean. Dean Brewer, a friend of LeMond's, traveled with him on the 1999 Tour. "At that time Greg was actually a big fan of Lance. We met Lance several times during the race, at the start on different mornings. Greg would give Lance advice and he was rooting for him."

Perhaps the first seeds of doubt were planted at a classy restaurant on a summer's evening in Provence, deep in the south of France in 2000. LeMond was there at the invitation of Roger Zannier, sponsor of the team that helped the American win his third Tour. Seventeen years had passed since LeMond knew Zannier as chairman and managing director of the French children's clothing company Z, also known as Groupe Zannier. Like many compatriots, Zannier loved the sport of cycling and wished to sponsor a professional team. He chose the former Peugeot team since it was losing its sponsor, rebranded it, and called it Z. He was prepared to invest $2 million in an average team but preferred to put $4 million into a team that could win the Tour de France. In September 1989, he signed Greg LeMond: $5.5 million over three years for the rider he believed would win him the Tour. Ten months later LeMond delivered and, you could say, Zannier was the last Frenchman to win the Tour de France. Since then, the race has been dominated by foreign riders and backed by foreign sponsors. So pleased was Zannier by LeMond's victory that ten years later, he reassembled the riders and the team's support staff for an anniversary dinner in Provence.

The timing was, of course, carefully planned because the reunion took place at the same time as the Tour de France passed through Provence. LeMond was happy to accept Zannier's invitation because he's lived in Minnesota since 1994 and he rarely gets to hook up with his old teammates. And when you've been a champion, it's nice that once in a while someone remembers. LeMond took his family to Provence for the reunion. He was looking forward to seeing old faces—perhaps most of all that of his old Belgian mechanic Julien DeVriese. DeVriese is the father of all mechanics; the gent from Ghent who looked after Eddy Merckx's bike in the seventies, Greg LeMond's in the late eighties and early nineties, and Lance Armstrong's in the late nineties, through to the early years of the new millennium. LeMond had always liked Julien, and when LeMond returned to the United States for the last time in 1994, he bequeathed Julien his old Mercedes.

The mechanic normally would have been with Armstrong on the Tour de France, but that summer there had been a falling out and Lance had asked an American mechanic, Dave Lettieri, to work with him. DeVriese was needed only on the evenings before the race's time trials and had no trouble getting the time off for the reunion. On the night after the reunion banquet, LeMond and his wife, Kathy, went to dinner with Julien, his wife, Vera, and their son Stefan; an evening they also shared with friends of the LeMonds'—Jorge Jasson, Dean Brewer, and Emily Kelley. Because they'd spent almost eight years working together, LeMond and DeVriese had a lot of catching up to do and no one was surprised that they sat together.

In the way old-timers do, DeVriese complained that the sport was no longer what it used to be and said working for Lance was tough. The previous month he had been at a training camp in the Pyrenees with Armstrong, Tyler Hamilton, and Kevin Livingston, a rendezvous also attended by Dr. Michele Ferrari. DeVriese was not impressed by what he saw at that camp. According to LeMond, DeVriese said the Postal team was using a lot of drugs. DeVriese would later deny he said this. Dean Brewer was at the other end of the table and would occasionally tune in to the conversation between LeMond and DeVriese. "I heard the part about Lance and Kevin Livingston and Tyler Hamilton being on a training camp in the Pyrenees and how secretive the whole thing had been. . . . Later that evening, after Julien had gone, Greg talked about what he had heard from Julien, and I know he was rather upset about it. Greg's feelings about Lance were being changed by what he was hearing."

Jorge Jasson, who sat closer, has a similar recollection. "The tone of the conversation was that Julien and his son were saying how the atmosphere on the [U.S. Postal Service] team was so unfriendly now, how everything was being done behind closed doors, how nobody enjoyed being around Armstrong, and how the whole thing revolved around keeping doping secret and keeping secret what he was doing. That I remember very well."

LeMond came away from the evening with DeVriese with doubts about the Postal team, and another meeting with his former mechanic the following April made him even more suspicious. It was the week between the Paris-Roubaix classic and the Tour of Flanders, and DeVriese invited LeMond to eat at his son's restaurant in Ghent. They were driving to the restaurant when they got talking about the French police investigation into the Postal team. LeMond wondered how it was going and DeVriese said it would come to nothing because he had signed an affidavit saying the team carried the controversial drug Actovegin to treat his diabetes. According to LeMond, DeVriese indicated this was not the reason the team carried Actovegin and that he had, in fact, made a false affidavit. DeVriese later denied this. That same month, April 2001, the orthopedic surgeon Brian Halpern invited LeMond to be a guest speaker at a symposium for sports doctors in San Antonio. LeMond was attracted by the suggested theme of his address, the role of doctors in professional cycling teams, and he warned Halpern that the audience might not like his take on this subject. In LeMond's view, cycling teams hired doctors to help them with their doping programs and that was the principal reason for having them on teams. Warned about the thrust of LeMond's talk, Halpern felt even better about having him speak. As it turned out, LeMond tempered his view slightly, admitted that not every doctor who worked in cycling was a doper but emphasized that since doctors had come into the sport, doping had increased. Cycling was a sport with few injuries and if a rider got sick during a race, there was normally a doctor present, paid for by the organizers.

But it wasn't his own contribution that most animated Greg LeMond at San Antonio. The man who spoke directly before him, Professor Ed Coyle from the University of Texas at Austin, spoke about ergogenic aids and the role of supplements in improving athletic performance. Then Professor Coyle spoke about his work with Lance Armstrong, whom he had tested over a seven-year period at the University of Texas. What LeMond heard Coyle say was that Lance's improvement postcancer could not really be attributed to a change in his VO_2 Max—that is, his maximal rate of oxygen consumption—because it hadn't changed significantly. Neither could the improvement be attributed to his weight loss because that had been seven to nine pounds and not the reported twenty. No, the key component for Coyle was mechanical efficiency, or more simply, Armstrong's more efficient pedaling style. LeMond couldn't buy this. He had worked with a very good physiologist in Europe, Adrie Van

Diemen from the Netherlands, who argued endlessly that a more efficient pedaling style helped a rider's endurance. To bolster his argument, Vie Diemen spent much time working with riders prepared to change to a more economic pedaling style and was convinced they could improve by a half to 1 percent. If LeMond was hearing right, Professor Coyle was claiming an 8 percent gain for Armstrong from pedaling more efficiently.

LeMond thought of the ten-thousand-meter athlete who finds the perfect running rhythm, which helps him to move faster, but as he does, the demands on his oxygen consumption increase and if he continues at the higher tempo, he ends up short of oxygen. The same principle applied to cyclists. LeMond approached Coyle afterward and they spoke about the issues raised in the professor's address. There was no meeting of minds and LeMond returned home to Minnesota, his skepticism about Lance Armstrong reinforced.

In July of that year, 2001, a journalist called and asked LeMond for his reaction to the news that Armstrong was working with Michele Ferrari. There was a time when LeMond would have sought diplomatic immunity and not answered a question like that. Perhaps he now felt he knew too much. What is certain is that he believed he had to publicly condemn Armstrong's association with Ferrari. Any other response would have amounted to moral cowardice.

To Greg LeMond, it seemed things were happening for a reason. Unbidden, so much information had come his way: Julien DeVriese telling him at the Team Z reunion in Provence that all was not aboveboard in the Postal team, the meeting with DeVriese the following spring in Belgium and more clues to the team's questionable methods. Along the way, he heard about the alleged hospital room confession, and finally, there was the news of Armstrong's relationship with the notorious Ferrari. When he put it all together, LeMond found himself among the doubters. It wasn't a conclusion at which he wanted to arrive but he was led to it. When the story of Armstrong and Ferrari broke in July 2001, LeMond felt relieved that some of what he knew was in the open. That it was a front-page story in many European countries was not a surprise because Ferrari was due to stand trial on doping charges two months later.

Armstrong downplayed his relationship with the doctor, describing him as someone he was occasionally tested by, and this only because his true coach, Chris Carmichael, was based in the United States and unable to be in Europe as often as was needed. In any case, he was using Ferrari's expertise specifically to prepare for an attempt on the world hour record. There were knowing nods among seasoned cycling writers at this explanation because many of them simply did not believe Armstrong would attempt the record. And he never did.

Daringly, Armstrong claimed he had never tried to hide his relationship with Ferrari and that a number of journalists knew about it. That may have been, but the unnamed journalists didn't write what they knew and they certainly didn't claim prior knowledge when the story broke. Certain members of Armstrong's own team did not know he was working with Ferrari, and in 2000, when Armstrong produced his acclaimed autobiography, *It's Not About the Bike*, there was not one mention of his friend and trainer/doctor.

But when that journalist called and asked LeMond for his reaction to the news, cycling's law of silence was about to be breached again. He told the journalist how disappointed he was that Armstrong should be associated with Ferrari and that it was because of doctors like Ferrari that cycling's image was so bad. LeMond was asked how he viewed Armstrong. Did LeMond believe he was clean or did he suspect he was using drugs? Not wanting to make direct allegations, LeMond settled on a response that would be reproduced in countless languages all over the world. "If it is true," he said of Armstrong's story, "it is the greatest comeback in the history of sport. If it is not, it is the greatest fraud." That was as far as he was prepared to go, but it was far enough. America's three-time winner of the Tour de France was indirectly questioning the integrity of his successor, a man who had survived cancer and who was two weeks from winning his third consecutive Tour de France.

Of one thing LeMond was sure: he would hear from Armstrong.

Toward the end of July, two weeks after publicly expressing doubt about Armstrong, LeMond traveled to London to meet representatives of the multinational oil company Conoco, who were interested in sponsoring a cycling team. He returned on a direct flight from London to the international airport at Minneapolis–Saint Paul and was met by his wife, Kathy. As he climbed into the driver's seat of Kathy's Audi station wagon, his cell phone rang. Realizing who it was, he mouthed "It's Lance" to his wife as he answered. After he said hello and asked Armstrong where he was, the conversation got more serious. "Greg, I thought we were friends," Armstrong said in a way that indicated his displeasure. LeMond tried to pass off the rebuke by saying he, too, thought they were friends. Armstrong then asked if the comments attributed to him in the press were accurate. LeMond said they were. He had a problem with Michele Ferrari and was disappointed Armstrong was working with him.

They tell different stories about the conversation that followed. Armstrong said it was clear to him that LeMond was speaking under the influence of alcohol, because of the way he started yelling once their conversation became fractious. LeMond says he had just gotten off a flight from London and his wife had handed him the car keys to drive them both to their home, so he certainly had not been drinking. LeMond says Armstrong threatened him by saying he could get ten people to say he used r-EPO. He also claimed Armstrong said everyone used r-EPO. Armstrong says this is illogical: why would he call up someone to complain about a drug allegation and then say everyone used drugs? Armstrong also says he reminded LeMond that in Professor Sandro Donati's report into doping in cycling, produced in the midnineties, LeMond was mentioned as being involved with the Belgian doctor Yvan Van Mol.

LeMond said Van Mol was doctor at the ADR team when he was there in 1989, and the only treatment he received from him was iron supplementation. LeMond also said that he won the Tour de France in 1986, before r-EPO entered cycling, and that it was after its arrival in the sport that he couldn't compete anymore. What is certain is that it was a venomous phone call between America's two greatest cyclists. Armstrong says it turned into a nightmare conversation.

Sitting in the passenger's seat, Kathy LeMond soon realized this was no ordinary telephone call. "Since it was Lance calling, I took notes of everything that was said by Greg during the call," says Kathy LeMond, "and then recapped with Greg what Lance had said. Some of Lance's words I could hear because he was so loud. It was clear their conversation was strained. Afterward I pieced together the principal elements of what was said between them." Kathy LeMond still has the notes of the conversations between her husband and Lance Armstrong, and the following is an account of the key elements based on her notes.

LA: Greg, this is Lance.

GLM: Hi, Lance. What are you doing?

LA: I'm in New York.

GLM: Ah, okay.

LA: Greg, I thought we were friends.

GLM: I thought we were friends.

LA: Why did you say what you said?

GLM: About Ferrari? Well, I have a problem with Ferrari. I'm disappointed you are seeing someone like Ferrari. I have a personal issue with Ferrari and doctors like him. I feel my career was cut short. I saw a teammate die. I saw the devastation of innocent riders losing their careers. I don't like what has become of our sport.

LA: Oh, come on, now. You're telling me you've never done EPO?

GLM: Why would you say I did EPO?

LA: Come on, everyone's done EPO.

GLM: Why do you think I did it?

LA: Well, your comeback in '89 was so spectacular. Mine [his victory in the 1999 Tour de France] was a miracle, yours was a miracle. You couldn't have been as strong as you were in '89 without EPO.

GLM: Listen, Lance, before EPO was ever in cycling, I won the Tour de France. First time I was in the Tour, I was third [in 1984]; second time I should have won but was held back by my team [second in 1985, behind teammate Bernard Hinault]. Third time I won it [1986]. It is not because of EPO that I won the Tour—my hematocrit was never more than forty-five—but because I had a VO_2 Max of ninety-five. Yours was eighty-two. Tell me one person who said I did EPO.

LA: Everyone knows it.

GLM: Are you threatening me?

LA: If you want to throw stones, I will throw stones.

GLM: So you are threatening me? Listen, Lance, I know a lot about physiology; no amount of training can transform an athlete with a VO_2 Max of eighty-two into one with a VO_2 Max of ninety-five, and you have ridden faster than I did.

LA: I could find at least ten people who will say you did EPO. Ten people who would come forward.

GLM: That's impossible. I know I never did that. Nobody can say I have. If I had taken EPO, my hematocrit value would have exceeded forty-five. It never did. I could produce all my blood parameters to prove my hematocrit level never rose above forty-five. And if I have this accusation leveled against me, I will know it came from you.

LA: You shouldn't have said what you did. It wasn't right.

GLM: I try to avoid speaking to journalists. David Walsh called me. He knew about your relations with Ferrari. What should I have said? No comment? I'm not that sort of person. Then a journalist from *Sports Illustrated* called me. I've spoken to two journalists in total. Maybe I shouldn't have spoken to them but I only told them the truth.

LA: I thought there was respect between us.

GLM: So did I. Listen, Lance, I tried to warn you about Ferrari. This guy's trial is opening in September [2001]. What he did in the 1990s changed riders. You should get away from him. How do you think I should have reacted?

That phone call wasn't the only one LeMond received concerning his publicly stated views on the Armstrong-Ferrari alliance. Within days, Thom Weisel called. Weisel's approach was conciliatory, as he told LeMond how he wished he had been around to sponsor him during his great years. LeMond didn't feel this was the reason for a call from a man whom he had not seen for three years and from whom he had never received a phone call. About what LeMond had publicly said concerning Armstrong, Weisel merely said he didn't think it was good for Greg to say those things. LeMond felt he was being told to be mindful of what he was saying because he might be the one to suffer in the end.

Perhaps if Weisel's had been the only such call, LeMond would have passed it off as a well-intentioned attempt to maintain harmony in the U.S. cycling family. But then a friend of LeMond's, Terry Lee, the CEO of the bicycling accessories company Bell Helmets, called. Lee's approach was friendly but the message was unmistakable. He didn't think LeMond should be the guy saying what he was saying. To Weisel and Lee, LeMond made the point that he believed Lance should not be working with Ferrari, whom he considered a dirty doctor, and he told them he felt entitled to say this. But, Lee counseled, it wasn't doing LeMond any good, and if he (Lee) were in the same position, he wouldn't do it. John Bucksbaum called and offered similar advice. CEO of General Growth Properties, a real estate investment company employing forty-seven hundred people in the United States, Bucksbaum has long been a cycling fan and he had twice traveled to the Tour de France with LeMond. He said he was ringing as a friend. Toward the end of a week of phone calls, LeMond received a message either by e-mail or voice mail, he can't remember which, to call John Burke, owner and chief executive officer of Trek, the bicycle company that has a licensing agreement with LeMond to manufacture, market, and distribute LeMond bicycles.

Burke was in a difficult position because his company also sponsored Lance Armstrong and the U.S. Postal Service team. According to LeMond, the Trek boss said he was under pressure from Armstrong and his attorney Bill Stapleton to do something about what he (LeMond) had said. Burke needed LeMond to publicly retract his earlier statement about Armstrong and Ferrari, and if that happened, it would be possible for Trek and LeMond to continue working together. If that retraction didn't materialize, LeMond would be jeopardizing his business because it was possible that Trek would have to end its contract with LeMond Racing Cycles. By now LeMond's head was spinning with the advice, counseling, and veiled threats of four of the most influential figures in the U.S. bike industry.

"It was like the troops were mobilized to shut Greg up," says LeMond's wife, Kathy.

The pressure took its toll on LeMond, who was worried about the plight of his company LeMond Racing Cycles. Burke said there was a clause in LeMond's contract with the company that said if LeMond did anything to hurt Trek's interest, that invalidated the contract. Criticism of Armstrong did not help Trek. LeMond's attorney, Chris Madel, tried to broker a settlement but Armstrong's people insisted on a retraction. There were so many phone calls, so many attempts to find a wording that might work, that in the end, LeMond grew so weary he told his lawyer to do whatever was necessary to bring some kind of closure to what had become a complete mess. In mid-August 2001, a month after LeMond's comments about Armstrong, an apology from LeMond appeared in *USA Today* and soon circulated through the cycling community.

It stated:

> I sincerely regret that some of my remarks . . . seemed to question the veracity of Lance's performances. I want to be clear that I believe Lance to be a great champion and I do not believe, in any way, that he has ever used any performance-enhancing substances. I believe his performances are the result of the same hard work, dedication and focus that were mine 10 years before.

Contacted by Sal Ruibal, the journalist to whom the statement had been given, Armstrong expressed his satisfaction. "It is nice," he said, "to hear there was a clarification. I've always had a lot of respect for Greg as a rider and for what he's done for our sport. I respect and appreciate him even more for going out of his way to say that. I didn't have hard feelings before he made the statement and don't have them now." Asked whether the statement might have been the result of pressure applied by Trek, Armstrong said, "If there was pressure, it was from cycling fans, his and mine, who were confused and felt the remarks were somewhat bitter."

The first time LeMond saw the statement that carried his name was in that day's copy of *USA Today*. Seeing it in print sickened him. It did not represent how he felt, it wasn't what he believed, and it was not what he said. From a controversy that cost him a lot of money in legal fees, that consumed so much time and caused no little amount of emotional stress, his great regret is that his so-called apology saw the light of day.

Chapter 15

THE STRANGE WORLD OF TY

At the time, you might have found three Americans who didn't care for Mom's apple pie, maybe two who didn't believe in the flag, but you wouldn't have found one who disliked Tyler Hamilton. A bike racer by profession, a saint by inclination, he was soft-spoken, impeccably well mannered, naturally gentle, and had the knowing air of a young man who had been here before. He was also talented, gritty, successful, and innately humble. Certainly, if ever he chanced upon a geyser, he wouldn't have come back and told you he'd invented hot water. Or so it seemed. Part of it was his apparently uncomplicated nature, but it also had to do with his starting point in life; you didn't leave Marblehead, Massachusetts, with airs and graces. Andrew James, a New Zealand mechanic with the Phonak team, for whom Hamilton rode in 2004, remembers Tyler's thoughtfulness. "At our first meeting, we were sitting down, and even though he was the star, I found myself talking about New Zealand and how I came to be living and working in Switzerland. Tyler had a way of asking questions about you, always turning the attention away from himself, and you didn't find many pro cyclists who did that."

Emma O'Reilly spent five years with the U.S. Postal Service team, and as Hamilton was there throughout her time, she knew him well. Of all the riders she worked with, Hamilton was the only she could have fallen for. And it wasn't his looks but the understated decency. In his engaging book *Lance Armstrong's War,* Dan Coyle writes about Hamilton in a chapter titled "The Nicest Guy." "He was extraordinarily polite," observed Coyle. "He had an awareness of small needs, his hands reaching out to open doors, tickle babies, pay for coffees, scratch exactly the right spot behind a dog's ears. He was, in short, the living image of the nicest guy you've ever met." The journalists who turned up at Hamilton's home were welcomed inside because he was old-fashioned that way; at his place, you were the guest. With their notebooks full of nice quotes, the reporters went away with the jaunty air of having had an audience with one of life's gentlemen. Of course, being Mr. Nice meant he didn't offend anyone; whatever the question, Hamilton pretty much always found a sweet answer.

No one complained because at the peak of his career, Hamilton was a good story. Not a serial winner like Armstrong, but a gallant challenger prepared to go through God knows what to get to the finish. He was second in the 2002 Giro d'Italia even though his shoulder was broken. The pain was so excruciating he ground his teeth down to the roots, so badly that twelve of them had to be rebuilt when he returned to America. If that was bad, the 2003 Tour de France was worse (or better, if you were a journalist reporting the story), because in the very first road stage of the race, some silly French rider fell, causing chaos among those in his slipstream. Blobs of blood and bits of human flesh littered the road, but it was Hamilton who fractured his right collarbone and then refused to quit. For three weeks, he rode through the pain, again grinding away those capped teeth, this dentist's dream making a miraculous breakaway to win the race's final mountain leg. Before they got to Paris on the final Sunday, he had climbed to fourth place overall. And it wasn't just the never-say-die spirit but the way it was accompanied by chivalry: when Armstrong crashed on the ascent to the Pyrenean ski station at Luz Ardiden during that 2003 Tour and rivals weren't sure whether to attack or wait, it was Hamilton who slowed them, who said that in a race as cruel as the Tour, the rule about not profiting from the race leader's misfortune must remain sacrosanct. In moments of triumph, his class showed. Inside the last kilometer of his Tour stage victory, he decelerated and, in front of the television cameras, waited for his CSC team car to pull alongside and warmly shook the hand of his directeur sportif, Bjarne Riis. Hardly had he crossed the line when he dedicated the win to his teammates. That was Tyler.

He struck a chord with fans, inspiring Lance Armstrong's ghost writer, Sally Jenkins, to light up the pages of *The Washington Post* with a heartwarming tribute:

> If you're sick of bad-boy athlete stories, of greedy, entitled creeps and ruined angels and two-faced paragons, and you're in need of a restorative example, turn on the Tour de France and watch Tyler Hamilton ride through the mountains with a fractured collarbone, so determined to finish the race that he'll have to get his teeth recapped from clenching them against the pain. . . . The Tour is a peculiar event, and it's had its doping scandals in the recent past, but this week it represents something well worth examining, and that something is personified by Hamilton, the lesser-known 32-year-old American from Marblehead, Mass., who has continued riding out of honor—a gratuitous honor.
>
> If Hamilton—who won his first ever stage in cycling's premier race on Wednesday while grimacing in discomfort—gets through the rest of the Tour, which wraps up on Sunday, he will have cycled more than 2,000 miles, over a hundred miles a day for three straight weeks, despite the collarbone fractured in two places and a badly twisted spine. He's a one-

man antidote to the thieves' mentality so many pro athletes have—which is to say: "If I can get away with it, I'd be a chump not to."

The thing about Tyler was he could never be a two-faced paragon. He was the guy you wished your daughter might bring home; the steely, well-behaved young man who reminded us of the way we once were. The privilege of his intimate acquaintance fell to Haven Parchinski, who was working at an advertising agency in Boston when she met Hamilton at a cycling race in 1996. They were married two years later. At first they shared an apartment in Brookline. She then worked as an account executive with Hill Holiday and he raced in Europe. During the 1998 Tour de France, he would call mornings at seven forty-five, which happened to be one forty-five A.M. back in Boston, but Haven didn't complain. For a man like Ty (as she liked to call him), it was the least you would do. After they were married and went to live in Nice, France, and then Girona in Spain, the other cycling wives marveled at Haven's devotion. They could be having a good, girlie coffee or even a late lunch, but if Ty was due back from training, Haven couldn't hang around. Everything was dropped and made ready for the returning man. They all thought Ty a lucky man while pitying their own poor husbands.

Haven and Ty didn't have kids but that was by design; they decided to wait until his career was over, and in any case, they had Tugboat, the golden retriever who was as central to their existence as any animal could be. When they showed family photographs, Tugboat was at the center of every shot. Friends looked at Tyler's relationship with Tugboat and honestly wondered if he could have loved a child any more. Tyler was at the 2004 Tour de France when Tugboat, struck with cancer, became critically ill. When it was certain the poor dog would not live much longer, Haven Hamilton placed him in the car and drove from their home in Girona to Limoges in France, where Tyler and Tugboat slept side by side one last time and the master said goodbye to his faithful dog the following morning.

So when the news of Tyler Hamilton's positive drug test came in

September 2004, it was like hearing that the late Mother Teresa had been involved in a twenty-five-year affair with a married man or that Nelson Mandela had secretly collaborated with the white South African government during his years of imprisonment. Professional sport was riddled with cheats, of course it was, but Hamilton wasn't one of them. You could believe it about anybody else but not Tugboat's owner and friend. Cheating was what bad people did, not something Saint Ty would ever get involved in. The evidence was damning but impossible to believe.

One heartbroken fan, identifying himself as Matthew from Brussels, sent an e-mail to Hamilton's website. "Oh no. Say it ain't so, Tyler." Tyler would spend a lot of time doing just that.

The thing about Tyler Hamilton was that what you saw was not what you got.

Meticulously attentive to the details of good behavior, he was also fiercely ambitious. But nature had not been as generous to him as to other elite athletes. He just wasn't born with the natural resilience necessary to win big stages. "I always had questions about Tyler's recovery," said Emma O'Reilly. "That's just my observation. I worked with him and you could see him getting progressively weaker and weaker through a stage race. But I had the utmost admiration for him because he would just dig in and dig in. Every day he used to dig in more but with less reserves to dig into."

In the early years this was something his U.S. Postal Service teammates saw as well, the gradual depletion of his strength until he was too wasted or too sick to carry on. Neither was the fragility purely physical. Before important races, Hamilton got nervous, and they wondered if some of his many crashes were not caused by anxiety. When he left Postal at the end of 2001 to become a leader at the CSC team, they thought it was a good move for him but doubted he would be able to truly compete as a leader in a race such as the Tour de France.

That wasn't to deny his determination, and neither should it be forgotten that Hamilton's desire to succeed wasn't far removed from Armstrong's. Little things stayed in the mind of his teammates, like the time he decided to lose weight. Pencil-thin to start with, he didn't lose more weight easily. So he went for long training rides, five or six hours, and restricted himself to a few Diet Cokes and water—forsaking the carbohydrates taken on long training rides. At the same time he would go to the outdoor market near where he lived in Nice and buy the fresh vegetables that seemed to be his staple diet. Though his relationship with Armstrong was unfailingly polite, at least for as long as he stayed with the Postal team, everyone could see he aspired to be where Armstrong was. After one of the Tour wins, the Hamiltons and Armstrongs were eating at a restaurant in Nice when the patron began to make a fuss of Armstrong. Hamilton listened politely as the champ and the restaurateur shot the breeze, and he couldn't help noticing Armstrong had not introduced him to this new admirer. Armstrong wasn't the only one who carried a little black book inside his head. On the team they thought Tyler was last in the line of people you would suspect of doping, until you realized how competitive he was—then he leapfrogged right to the front.

Dr. Prentice Steffen, the first doctor employed by the Postal team, has talked about a time when he was approached by Hamilton and another Postal rider, Marty Jemison, at the Tour de Suisse in June 1996. According to Steffen, they couched the request in euphemistic terms but his understanding was they wanted him to provide the team with doping products. When the allegation was publicly aired, Hamilton dismissed it, saying Steffen was mad because he was fired and had made up the story. There was no explanation as to why Dr. Steffen would include Hamilton, a rider he liked, in his concocted story. If that raised a question about Hamilton, it still didn't cause a ripple on the smooth surface of the reputation of a man who was considered perfectly clean. But for those watching closely, there were signs that caused one to wonder.

Riding with the Postal team, Hamilton was introduced to Dr. Michele Ferrari and was impressed by what he saw. At first he seemed to tag along with Armstrong and Kevin Livingston, who had formal relationships with the Italian doctor, but after a while Hamilton became a full member of the Ferrari team. Why should Frankie Andreu say no to Ferrari and Hamilton say yes? Why would Jonathan Vaughters, an excellent climber on the Postal team, not even know Ferrari worked with three of his teammates? Ferrari's fees were exorbitant but one heard nothing but praise for his work. Though he was well paid anyway, Ferrari received an expensive Rolex watch from Armstrong, Livingston, and Hamilton after the 2000 Tour de France. Once he left Postal at the end of 2001, Hamilton's days with Ferrari were over because the doctor's arrangement with Armstrong prevented him from working with any of his Tour de France rivals.

Losing out on Ferrari's expertise was offset by Hamilton's linking up with Dr. Luigi Cecchini after joining the CSC team in 2002. Cecchini had spent a year working at the University of Ferrara in his formative years and became friends with Michele Ferrari. As well as sharing an interest in physiology and the science of helping sportsmen perform better, Ferrari and Cecchini were also keen cyclists, and their training spins tended to be competitive and won by the latter. In the midnineties, the height of the r-EPO era, they were the two dominant trainers/doctors of the time. Cecchini's star pupil was Bjarne Riis, the winner of the 1996 Tour de France and the rider known as Mr. Sixty Percent because of the perception that he raced with a hematocrit of an extraordinary sixty.

Compare with Hamilton's the path taken by Vaughters after he left the Postal team in 2000. He joined the French-based Crédit Agricole and enjoyed the team's minimalist approach to medication. The idea of working with an Italian doctor with a questionable reputation wouldn't have occurred to Vaughters. But his results remained decidedly ordinary, whereas Hamilton climbed to a new level after leaving the Postal team. In his first year at CSC, 2002, he finished second in the Giro d'Italia, the following year his fourth place in the Tour de France was the subplot that almost transcended Armstrong's feat in winning his record-equaling fifth Tour. Hamilton also won the Tour of Romandie twice and also the prestigious one-day classic Liège-Bastogne-Liège. Not only did he prove himself capable of leading a team but his performance at CSC earned him an improved contract offer from the rival Phonak team toward the end of the 2003 season. Phonak promised Hamilton absolute leadership of the team and a say in the management.

Finally, Hamilton's position in the peloton was getting close to Armstrong's; they each worked with an infamously successful Italian doctor, they were each absolute leaders of their teams with a strong input into how the team was run. Furthermore, when Hamilton compared his Phonak squad with Armstrong's Postal squad, he thought his team could be as good as his rival's. All through the years at Postal, this was the opportunity he craved—absolute leadership, the support of a strong team, the belief of the people who owned the team, the prospect of taking on his greatest rival in the 2004 Tour de France. For the first time, he thought, he would start from the same point as Armstrong.

That was when it all went wrong.

The fight against doping in sport was once likened to the individual pursuit in track cycling, where the two competitors start at opposite sides of the arena and once the starter's gun is fired, the aim is to gain ground on the other guy. On one side, there is the athlete who dopes, on the other side the tester trying to catch him. Since the race began however many years ago, the doper has stayed comfortably ahead of the tester. A good example of how the race plays was the response of cyclists to the discovery and validation of a test to detect r-EPO in 2000. As it was the drug of choice for so many, the new test forced riders to change their ways. At first they switched to micro-doses of the drug, making it harder for the testers to detect it; when the testing became more sophisticated, the riders used undetectable blood transfusions. Mostly, this involved extracting and later reinfusing their own blood. Finnish athletes benefited from this method of doping in the 1970s; the USA cycling team at the 1984 Los Angeles Olympics also used it; and under the direction of Professor Francesco Conconi, Italian distance runners transfused their blood throughout the eighties. Conconi and his protégé Michele Ferrari also worked with Francesco Moser when the Italian cyclist blood-doped to set a new world hour record on a track in Mexico City in 1984.

Though no one disputed the performance-enhancing effects of transfusing blood, it wasn't as straightforward as injecting r-EPO. Because the process was labor intensive and involved trained medical people in a time-consuming, unethical, and illegal practice, it was also far more expensive. Over the last five years, doping at the highest level in cycling has become sophisticated and the preserve of those with the financial wherewithal to spend anything between $60,000 and $100,000 on a season-long program. That constitutes the annual salary of a regular équipier and stops him from even considering blood transfusions. Doping became a rich man's sport. The switch from r-EPO to transfusions did not go unnoticed by anti-doping authorities because they continued to see biological values indicating blood manipulation but without any evidence of r-EPO. The difficulty, though, was that transfusions were undetectable.

Actually, that wasn't completely true. Autotransfusions, where an athlete extracts and reinfuses his own blood, are undetectable, but in 2002 the anti-doping body Science and Industry Against Blood Doping (SIAB) set about finding a test for homologous transfusions, where an athlete transfuses someone else's blood. "We identified it as a risk, something athletes could be doing," said Mike Ashenden, project coordinator for SIAB. "We sat back and thought, 'Well, who could tell us how to detect mixed cell populations?' Margaret Nelson is at the Royal Prince Alfred Hospital in Sydney and she had already been using a similar method to detect when a baby's red cells leak into the mother's bloodstream. So we took that method, modified it, using more antigens than she needed for her test. We then published our results, demonstrating that the test could be performed in other labs with different people performing them."

The United States Anti-Doping Agency (USADA) funded SIAB's research and when the work was complete, it was passed on to the World Anti-Doping Agency (WADA), who took charge and asked SIAB not to publicize the fact that there was now a test for homologous transfusions (involving blood from someone else with the same blood type). During the process of validating the test, SIAB asked a number of laboratories in the Sydney area to perform the test, using their own instruments, and also gave the method to the IOC-approved laboratory in Lausanne, which does much of the UCI's anti-doping testing. To practice its use of the test, the Lausanne laboratory used blood samples collected from professional cyclists. One of the samples showed two cell populations, a clear indication the rider had transfused someone else's blood. The rider with mixed cell populations was Tyler Hamilton.

Sometime before this, in late April 2004, the UCI had good reason to be suspicious of Hamilton. On April 24 he was blood-tested before the Liège-Bastogne-Liège classic in Belgium and his blood parameters, namely his high hemoglobin and low level of young red blood cells (reticulocytes), indicated doping. If an athlete uses r-EPO or has a blood transfusion, this will increase his hemoglobin and his body, reacting to the increased hemoglobin, naturally stops producing its own new red cells. And so the combination of high hemoglobin and low reticulocytes is, from the testers' point of view, highly suspicious. Under a mathematical formula worked out at the Australian Institute of Sport (AIS), a number, called an "off-score," was given to this relationship between hemoglobin and reticulocytes. Clean athletes record an off-score around 90; those who have used r-EPO might have a score of 130 or greater. According to the UCI's health test rules, any rider with a score of 133 or more will be withdrawn from competition until his blood values drop to beneath the 133 threshold.

From that April 24 test, Hamilton's off-score was 123.8, unusually high but not above the threshold. He was again blood-tested five days later, before the start of the Tour of Romandie, which he would go on to win, and his off-score was 132.9, fractionally below the permitted number. At the time, his hematocrit was 49.7, again just beneath the permitted limit of 50. As well as these highly suspicious values, the Lausanne laboratory also detected mixed cell populations, demonstrating Hamilton had someone else's blood in his body. The governing body sent the rider a letter saying his values showed strong signs of a "possible manipulation." Around this time Mario Zorzoli, medical director of the UCI, met with Hamilton and showed him a histogram, the diagram that shows one red cell population that is your own and a second that belongs to someone else. Zorzoli's area of expertise is blood and he would have fully understood the clarity of the result.

As an anti-doping administrator, Zorzoli was in a difficult position. Hamilton's off-score number and hematocrit on April 29 strongly suggested doping and the mixed population demonstrated it, but neither set of results constituted a standard positive drug test. The off-score and hematocrit results came from health checks, and while they had gone right to the limit, they did not exceed it. Though the new test for homologous blood doping showed a clear case of manipulation, the test would not be officially sanctioned until the start of the Olympic Games two months later. What was Zorzoli to do? Hamilton could not be sanctioned and yet the authorities knew he had used someone else's blood. A few days after his meeting with Zorzoli, Hamilton did another blood test and produced significantly lower values than the April 29 test. One possible explanation for this is that Zorzoli insisted he get his blood values down and much closer to normal, which could have been achieved by extracting blood.

Hamilton started the Tour de France a few weeks later with mixed cell populations but was saved by the fact that the new test would not be official until the Athens Olympics. That made little difference in the end because Hamilton's tour was over before it truly began; a crash on stage six to Angers resulted in a back injury from which he did not recover. He carried on for another week but the injury affected him badly in the mountains and he abandoned on stage thirteen to La Mongie in the Pyrenees. Though it was a major setback to have to quit, Hamilton allowed his body to recover and then started his preparation for the Athens Olympics, where he intended to compete for the United States in both the road race and the individual time trial. Did he know going there that the test for homologous blood doping would be in place? If so, did he mistakenly believe his blood would no longer be showing evidence of foreign red cells?

If it is questionable what Hamilton's mind-set was, Mario Zorzoli's thinking was more unfathomable. The UCI medical director knew Hamilton had a mixed cell population and knew also that the means to officially detect it was now available. "I had a lot of sympathy and empathy for the UCI until this case," says Mike Ashenden, who had led the team that came up with the new test. "I can understand why they would be reluctant to embroil themselves in controversy and bust too many cyclists. But when you know somebody has done this [blood doping], and you don't go after them, there's no excuse in my mind for that. At that point, I was disgruntled, to say the least, with their attitude. We bust our guts to get the test in place and they then don't see it through."

As happened at the Sydney Olympics four years before, the U.S. cycling team did not stay with the vast majority of their compatriots in the Athletes' Village but in separate housing away from the village. Within USA Cycling there were some raised eyebrows at this because it was felt housing outside the village left the team vulnerable to visits from unauthorized personnel. Why should the cyclists be different from the other U.S. teams, when even the superstars of the NBA were happy to be accommodated in the Athletes' Village? Hamilton's victory in the individual time trial on August 18 ensured he would be tested; the following day, his blood was analyzed. Dr. George Paterakis, the Greek hematologist hired by the Athens laboratory to examine blood samples, told the *Chicago Tribune* what he found. "It was a crystal clear case of blood doping," he said of sample number 6825, which he would later learn belonged to Hamilton. Paterakis reported the case to Costas Georgakopoulos, the director of the Athens laboratory. Georgakopoulos decided the sample was not positive but suspicious, and as such, it was deemed negative. In line with the IOC's protocol, the second half of sample 6825, known as the B sample, was frozen.

Three weeks later, the head of the IOC's medical commission, Arne Ljungqvist, received a batch of e-mails relating to sample 6825, one of which came from Paterakis, who had been very surprised to learn that the positive result for homologous blood doping had not been followed up. By now, there was a realization that the laboratory had made a serious mistake: the A part of sample 6825 was indeed positive, and the frozen B sample was sent to the Lausanne laboratory to see if enough cells survived to conduct the follow-up test. Not enough had, and without a proper B sample, the test result reverted to negative. Hamilton escaped not on a technicality but because a serious mistake was made at the Athens laboratory. When this tale of gross incompetence emerged the following month, reactions ranged from disappointment to downright disgust. The normally restrained Bobby Julich, the American cyclist who finished third in the time trial, was disgusted. Julich had been surprised that people he didn't know were visiting Hamilton, but Julich was not surprised when he was asked if he believed his roommate had doped. "It goes against everything I've ever seen or known from the guy. But the rest of us at the Olympics passed the test. Why didn't he? I'm sick of people who cheat, sick of cleaning up their mess and trying to explain it. There is heavy evidence against him. With that much evidence, I don't know how he's going to get out of it."

WADA and the IOC were angry that someone who, in their mind, had tested positive should have left the Olympics with a gold medal. But they knew the evidence of homologous blood doping didn't quickly disappear, and as Hamilton was riding the Vuelta a España two weeks later, there would be an opportunity to test him again. Their desire that he be tested was communicated to the UCI, and after Hamilton won a time trial on September 11, he was again asked for a blood sample. This time there would be no mistakes; samples A and B were both positive. On the evening of September 16, Hamilton was told the bad news by his Phonak directeur sportif, Alvaro Pino. To the Phonak director, Hamilton appeared remarkably calm and sure of his innocence. "I spoke with the rider, and knowing him as I do, I'm relatively calm. He told me, 'Be calm, because this will work out in my favor and I'm telling you that sincerely, because there's absolutely nothing in this.' "

Hamilton was equally certain during his first press conference after the announcement of the positive test, which was held at the

Phonak's team headquarters in Regensdorf, Switzerland. "I'm devastated to be here tonight," he said. "I've been accused of taking blood from another person. Anyone who knows me knows that is completely impossible. I can tell you what I did and did not put into my body. Cycling is very important to me, but not that important. If I ever had to do that [blood doping], I'd hang the bike on the rack." He carried on in that way, soft-spoken but passionate, a young man whose life had been torn apart by someone else's mistake and who wanted everyone to understand that this allegation of cheating besmirched his character and he would fight it to the end. "I am 100 percent innocent. I will fight this until I don't have one euro left in my pocket. I have always been an honest person."

European and U.S. newspapers carried reports of Hamilton's cry of innocence the following day, and in the midst of all the denials there was a short report in the Spanish daily newspaper *El Pais* that went largely unnoticed. Its key point was a quotation from the Phonak doctor Iñaki Arratibel. "Three days ago, when we received the official form from the UCI announcing the positive test, I phoned Hamilton, who by the way has his own doctor, the Italian Luigi Cecchini. He told me he was innocent, that he never had a transfusion to improve his performance, that the detection method was new, unproven, and insecure. He also said he was in Switzerland with his lawyers and that everything could have come from a surgical procedure that he had undergone a few years before, a procedure in which a transfusion was needed."

Arratibel was asked what kind of surgical procedure Hamilton had needed, and the doctor said that when he asked that, Tyler had not been able to explain himself clearly on this point. Hamilton never again mentioned the "surgical procedure" explanation he had given to Arratibel.

Hamilton's denial of the charge, however, remained vigorous. He welcomed the opportunity to make his case publicly, and in return for a fair hearing, he offered journalists a heartrending story of a good man wronged. Blood doping wasn't something a person like him would do. In an interview with *The Boston Globe*, he stressed how cheating went against everything he believed. "Everyone who knows me as a person knows I didn't do this. I grew up in a family where being honest was so, so important. Being called a cheater, knowing I didn't cheat, it's the worst feeling in the world. It's like somebody stabbed me in the back."

The difficult part for Hamilton and his high-profile attorney, Howard Jacobs, was taking the knife out. However often Hamilton said the test was new and suggested it was unreliable, the truth was that it was a well-established test—the only thing new was its application to sport. With regard to its reliability, the test had been used for many years in obstetrics to help safeguard the lives of mother and baby in childbirth. Hamilton based his defense on two points. One, the test did not have a threshold beyond which the number of foreign cells constitutes a positive. In other words, the mere presence of foreign cells, regardless of the amount, was enough to establish that an offense had been committed. Two, the test did not allow for "false positives." The example frequently quoted was the presence of cells from the possibility of a pregnancy that had started out as twins and ended up as a single baby. Before the twin is absorbed by the mother, some of its cells enter the body of the other fetus and remain there for life—the vanishing twin theory.

There were many who wanted to believe in Hamilton's innocence, their yearning eloquently expressed in e-mails of support sent to the rider's website. "I do not believe them," wrote Perry of Augsburg in Germany of Hamilton's accusers. "If you are a liar, every cycler does it." Chris Davenport, a friend of Hamilton's, set up a website, www.believetyler.org, and there was a supportive e-mail from Lance Armstrong. A fund was set up to raise money for Hamilton, perhaps to ensure that he wouldn't, after all, have to spend his "last euro" on his defense. Terry Lee, the CEO of Bell Helmets, aligned his company to the Hamilton cause by giving away I BELIEVE TYLER buttons, and Andy Rihs, chairman of Phonak, backed his man. "I don't believe in the test. I think this test was done sloppily. . . . I don't fire innocent people." Rihs's pledge of loyalty was delivered on September 21, 2004—two months before Phonak terminated Hamilton's contract.

As happens in doping conflicts, the case dragged on and was first heard by USADA in early March 2005. The three-man panel found Hamilton had committed a doping offense and gave him the statutory two-year ban from competition. Hamilton opted to appeal to the Court of Arbitration for Sport (CAS) and his case was retried in January 2006 at the Brown Palace Hotel in Denver. Again, the panel ruled against Hamilton. In the seventeen months since the announcement of the positive test at the Vuelta a España, Tyler Hamilton had based much of his defense on his personal integrity, and he continually suggested that people who knew him knew he would not cheat. This was a theme explored by Travis Tygart, counsel for USADA, in his cross-examination of Hamilton at the CAS appeal.

> **Q:** Tyler, you are an elite level athlete. You're used to having your name in the paper, having a perception out there about you. As an elite level athlete, do you think it's difficult sometimes to be completely forthright and honest in situations where you would be embarrassed or where there may be consequences?
>
> **A:** No, not really.
>
> **Q:** So you think it is easy in situations—even where you may be embarrassed—it's easy for you to be truthful.
>
> **A:** I always try to be truthful, that's the way I live my life.

Tygart then asked Hamilton to look at his sworn witness statement in which he explained why he pulled out of the 2004 Vuelta a España after learning of his positive dope test. To stay in the race would, he said, have been unfair on the other contenders for overall victory. But the USADA prosecutor also directed Hamilton to that part of the witness statement where he acknowledged that he told his fellow riders and the public his reason for quitting the race was an upset stomach.

> **Q:** Do you acknowledge that you told the public something different than the truth?
>
> **A:** Yes, although it was . . . I did have a stomach problem.

There was another example of a little white lie when Hamilton, suffering from a bladder infection during an early-season race in 2004, decided it would be embarrassing to have to tell journalists he was urinating blood, and instead told them he was suffering from a stomach infection. "Again," said Tygart in his cross-examination, "would you agree this is another incident or example when you told the public something other than the truth?"

> **A:** Yes. You are right, yes.
>
> **Q:** And the reason you did that was because you were worried about being embarrassed by the truth?
>
> **A:** Well, I mean, it wasn't . . . I guess you could say it was a massage of the truth because I did have some stomach problems.

Not short of examples of Hamilton's occasionally cavalier attitude toward the truth, Tygart referred to an interview Hamilton had done with John Henderson of *The Denver Post* in which he talked about his chances of winning his appeal to CAS. "Very optimistic," Hamilton had said in answer to a question about his prospects. "I thought our chances were good in the first hearing, although we were missing a lot of info, a lot of data on our own test results. Now that we have the data, we have a better case." Tygart went through Hamilton's sworn statement for the appeal, in which it was made perfectly clear there was no new data. His boast to Henderson was utterly without substance.

During the first hearing with USADA, Hamilton's college education was discussed. It is an interesting subject because he has often been portrayed as a man cut from a more refined cloth than the ordinary professional cyclist. In an interview with Matt Seaton of the English daily newspaper *The Guardian* in June 2004, the journalist said Hamilton was not a man to be underestimated. "He is, for a start, an economics graduate from the University of Colorado at Boulder." Seaton then quotes Hamilton on the subject of his own erudition: "For me, having the opportunity to go to college was very important. To miss out on an education is a loss. I feel like I was lucky that I didn't realise I was a good cyclist until later. You go to a race now and you go around and interview riders. I'd say 95% don't have education past high school."

Then, when Hamilton tested positive later that year, his impassioned plea of innocence drew on both his intelligence and his love of Haven. "Bike racing is my livelihood, it's very important for me, but it's not worth risking my life. They're accusing me of taking someone else's blood. Number one, I'd be risking my life. It could kill me. Number two, I could get a disease, AIDS, hepatitis, the list goes on. Number three, I could endanger Haven, and everyone knows how much I love her. . . . I'm not saying I'm a genius, but I have a college degree. I'm pretty smart. I don't take risks—I take *educated* risks. . . . This is a risk I would never take. If it were a life-or-death decision, okay, but like this, never."

Everything Ty said was plausible and reasonable, except the reference to his university career. Tyler did attend the University of Colorado but he did not graduate.

Chapter 16

THE EMPIRE STRIKES BACK

It is Tuesday, June 15, 2004. A number of high-powered business people have gathered in the Discovery Channel headquarters at One Discovery Place in Silver Spring, Maryland. In the midst of the corporate captains, Lance Armstrong stands out, for he is the athlete and the man around whom everyone gathers. Two and a half weeks remain before he will attempt to win a record sixth consecutive Tour de France, and though this get-together interferes with his preparation for the race, the rider knows the importance of the occasion. Tailwind Sports, the company that created and owns Armstrong's spectacularly successful team, needed a replacement sponsor for the U.S. Postal Service and came up with the Discovery Channel. For a three-year deal, beginning in 2005, the cable television network will pay over thirty million dollars and be the team's primary sponsor. So there are the sincere handshakes and upbeat talk, but there is also some foreboding because this is also the occasion on which Armstrong will first address issues raised in a controversial new French book, *L.A. Confidentiel: les secrets de Lance Armstrong*. Though plenty of journalists are present, their interest in Discovery's takeover of the team is secondary.

At the very moment when Armstrong and his new backers are breaking bread in Maryland, fresh copies of *L.A. Confidentiel* are being placed on European bookshelves, and its allegations are devouring space on the sports pages of newspapers across the continent. Coming so close to Armstrong's quest for a record sixth Tour de France victory, the accusations of doping are at best a serious irritation, at worst a catastrophe. Foremost among the rider's concerns is the need to reassure the new sponsor. So, Armstrong has come to Maryland with a plan. In fact, the operation had begun the previous day when his London attorneys, Schillings, issued a statement to media outlets advising them that

> Lance Armstrong has instructed his lawyers to immediately institute libel proceedings.
>
> 1. In the High Court in London against the *Sunday Times* seeking an injunction and substantial damages.
> 2. In Paris, against [the authors] David Walsh, Pierre Ballester, the publishers of *L.A. Confidential* and the publishers of *L'Express* magazine.
>
> Proceedings will be filed in the High Court tomorrow.

Sitting now in the offices of the Discovery corporation, Armstrong explains the reasoning behind the decision to sue. "This is not the first time I've lived through this," he said of the allegation that he has been involved in doping. "Every time, we've chosen to sit back and let it pass. But we've sort of reached a point where we really can't tolerate it anymore and we're sick and tired of these allegations and we're going to do everything we can to fight them. They're absolutely untrue.

"Enough is enough," he adds.

The press conference serves its purpose, as it enables Armstrong to reassure his new sponsor and the U.S. public that there is no reason for anyone to doubt the team's ethics. "We don't use doping products and we will sue those who suggest we do," he says. Judith McHale, president of Discovery Communications, has no doubts about the man who will spearhead the corporation's entry into the world of professional cycling. "Lance is a role model known for determination, integrity, and a spirit that never gives up. There is no better ambassador for quality and trusted information."

Suing the authors and publisher of *L.A. Confidentiel,* the publishers of *L'Express* magazine, and *The Sunday Times* wasn't the only litigation instigated by Armstrong. His legal team also tried to have an insertion in each copy of the book carrying his denial of the allegations. Judge Catherine Bezio heard the case on Friday, June 18, in Paris. Acting on Armstrong's behalf, Christian Charrière-Bournazel accused the authors of "dumping a load of garbage onto an immense champion." Representing the authors and publisher, Thibault de Montbrial argued Armstrong could have made his denials when the authors, over a four-week period, repeatedly tried to put questions to him. In his defense of the book, de Montbrial contended that investigative journalism could not be practiced if those under suspicion refused to answer questions and were then allowed to insert an all-embracing denial in what was being written about them.

Judge Bezio issued her judgment three days later. She found in favor of the book. Her verdict asserted that the allegations contained in the book "do not necessarily constitute defamation." Under French law no libel is committed if the allegations are made in good faith or if they turn out to be true. The judge also ruled that by seeking an emergency order, Armstrong's legal team had not allowed sufficient time for the court to consider the use of the good faith defense.

The court ordered Armstrong to pay costs and fined him eighteen hundred euro for what it considered an abuse of the legal process.

"By ruling against Mr. Armstrong, the judge perfectly understood that Armstrong has attempted to respond through the courts to the journalists' questions he previously refused to answer," said de Montbrial. "This decision guards the right of journalists to engage in serious investigations, and reminds us that a person who is the subject of such an investigation cannot use his refusal to answer questions as a means of getting a judge to impose censure." Armstrong's lawyer saw things differently. "I'm very upset and don't share the view of the court," said Charrière-Bournazel. He insisted he and Armstrong were not seeking the suppression of the book but an opportunity to insert a statement to readers expressing Armstrong's denial of the most serious charges. They appealed Judge Bézio's decision. Eleven days later, a Paris appeals court turned down Armstrong's request and *L.A. Confidentiel* continued to be sold without the insertion he sought.

It is two days before the start of the 2004 Tour de France and the great and good of the professional cycling circus are assembled in the Belgian city Liège. In the early afternoon, journalists file into the conference area at the press center for a scheduled rendezvous with the defending champion Armstrong. Since his emphatic denial of any wrongdoing at the Washington press conference sixteen days before, Armstrong has spoken little about the allegations contained in *L.A. Confidentiel,* but the time has come for a definitive statement.

One of the book's two authors is among the journalists in Liège. As he walks into the press center an hour before the Armstrong conference, the U.S. Postal Service directeur sportif Johan Bruyneel stands by the entrance speaking with some journalists. Noticing the book's author, Bruyneel shouts: "Hey, good job, good job." It is a taunt rather than a compliment, an expression of barely concealed hostility. In the conference area, most journalists are already seated and the sense of anticipation is palpable. Everyone wants to hear

Armstrong's reaction to *L.A. Confidentiel,* but there is the customary and almost surreal timidity one sees when sportswriters gather before a famous athlete: the journalists want the tough questions answered but no one wishes to ask them. This is especially so on this occasion and there is much tiptoeing around the issue before the relevant question is asked. Even then it comes coated in diplomacy.

"Lance, have the controversial allegations upset your preparation for the race?"

"No," he replies. "But I'll say one thing, since the esteemed author is here. In my view, extraordinary accusations must be followed up with extraordinary proof. The authors have had four or five years working on the book, and they've still no proof. But I will spend however long it takes and whatever it takes to show the allegations are unfounded. I have already engaged lawyers in England and France."

Marc Belinfante, a television journalist from the Netherlands, asks Armstrong how he feels about what his former teammate Stephen Swart said in the book.

"No comment."

Belinfante persists and asks about Emma O'Reilly's allegations.

Armstrong glares at Belinfante and says, "Next question."

Armstrong knew from summarized accounts of the book in various newspapers that Steve Swart, Prentice Steffen, Emma O'Reilly, and Greg LeMond were the four most important sources for the book. But someone had told him the authors had identified Betsy Andreu as another source, and that was interesting because her husband, Frankie, was a teammate for eight years and Armstrong found it hard to believe Frankie's wife would have come out against him—especially as Frankie still earned his living from cycling and was going to be working on the Tour de France for the Outdoor Life Network (OLN) television channel. Could Frankie, in his role as an OLN commentator, do his job if people at the U.S. Postal Service team decided not to cooperate with him? What would he say if they asked him about his wife? For sure, he would say his wife was not a source. But if that were the case, couldn't Betsy say publicly she had nothing to do with the book and that the authors were lying in suggesting she had? That would damage the book's credibility.

He discussed the idea with his attorney Bill Stapleton and they decided the best way forward was for Lance to speak with Frankie—man to man, friend to friend. It was the first or second day of the Tour and they got one of the team's soigneurs to make contact with Andreu. Could he drop around to the hotel for a little chat with Lance? They spoke about the publication of the book and Armstrong's information that Betsy was somehow one of the sources. Armstrong talked about the existence of a tape containing material from Betsy or even an affidavit from her that supported one or more of the book's allegations. Frankie Andreu was sure there was no tape, nor any affidavit, and he told Armstrong that. There was no rancour, no pointing of fingers; if Betsy wasn't a source, then maybe she could just come right out and say so. Easy. Andreu said she wasn't a source but avoided the question of whether she would publicly support Armstrong. Of course, Armstrong picked up on that and told his former teammate that Stapleton and Stapleton's colleague Bart Knaggs would talk to him later in the week to see what could be done.

In the evolving drama of Lance Armstrong and *L.A. Confidentiel*, Frankie Andreu wanted no part. What good could come of it? His living came from cycling, and, just as important, it was still his passion. He had retired from racing at the end of 2000, but the sport continued to be his life. Cycling wasn't perfect but what sport was? His wife was different. She put her moral values above everything—even, it seemed to him, above her family. Sure he could talk to her but he wasn't going to change her. When he explained to her that talking honestly about doping wasn't appreciated in cycling and if he did it too much, it would hurt his career, she told him to find another career. He didn't want another career, but as for Betsy coming right out and trashing the journalists, that wasn't going to happen.

Andreu's relationship with Armstrong was complex. They went

back a long way and had been close friends. After Andreu left the Postal team in 2000, there hadn't been much contact, but that was the same for all of the guys; once you left the team, you lost your place in Lance's world. But Andreu wasn't about to forget all the good times. Like when they shared the apartment in Como. That was in 1993. Lance liked that apartment, although he had to have a place by the lake the following year. "Why would I pay a thousand bucks a month for a place I'll hardly ever use?" Frankie asked when Lance wanted him to share the place by the lake. In the summer of '95 Betsy came to Como to see Frankie and she got to know Lance. There was one particular time when Frankie had to do an interview with the American network ABC, and Betsy and Lance went off to do the grocery shopping. They talked about relationships, marriage, where they had come from, and Lance talked about what it was like to grow up without a dad and how he had no wish to meet his biological father, who'd disappeared before Lance had had a chance to get to know him. Their conversation about religion was lively because Betsy is a devout Catholic and Lance is equally upfront about his atheism. His belief was that science offered an explanation for everything. Betsy asked him about those who made miraculous recoveries from cancer. "Good luck and good medicine," he said.

They stayed in Como until the end of 1996, when the worsening traffic spoiled the roads for cycling, and anyway, Lance and Frankie joined the French team Cofidis and it made more sense to live in Nice. By the beginning of 1997 Armstrong and his future wife, Kristin, were seeing each other, and during those early years in Nice, Frankie, Betsy, Lance, and Kristin saw a lot of one another. Lance loved Betsy's risotto and sometimes she would prepare it especially for him. Certain things weren't discussed, though. For example, what precisely was the nature of Lance's relationship with Michele Ferrari. Betsy had doubts, and was sure Kristin must have felt the same. There was the trip to the Milan–San Remo classic in March 1999. Lance and Kristin picked up Frankie and Betsy at their apartment in Nice and set off on the journey to Milan, where the race would start. On the way, Lance called the then governor of Texas, George W.

Bush, whom he wanted for a ceremonial role at his charity cycle, the Ride for the Roses. Bush wouldn't take the call, but Lance passed that off by saying he must have been busy. Lance and Betsy discussed politics and religion along the way. He believed in complete separation between church and state; she had a different view.

After about two and a half hours on the road, they pulled into an AGIP gas station by the side of the highway. "Do we need gas?" Betsy asked. "Nope," said Lance. "I need to see Michele." Betsy Andreu looked to her husband with raised eyebrows. "I didn't know about this," he said, quietly. They pulled up alongside a small recreation vehicle in the parking lot. "Come and get me in an hour," Lance said, as he jumped out of the car and headed over to Ferrari's camper van. Though she wasn't pleased with this hour-long stop, Betsy couldn't help noticing how cheerful Armstrong was as he left the car to meet Ferrari. *Giddy* was the word that came into her head. Kristin needed cash, so they drove to the nearest town, where Kristin found an ATM and took their dog, Boone, for a walk. They waited in the small hotel beside the gas station while Lance finished whatever he was doing with Ferrari in the camper van. Eventually Lance called Kristin, told her he was ready, and they made their way to the car. Ferrari came by to say hello, and when they set off, Lance gently chided Frankie about Ferrari. "You could get results, too, if you'd see him, but you're too cheap." Later, when Frankie and Betsy were alone, he said, "I sure as hell don't want to give him [Ferrari] 20 percent of my salary." After they left the gas station parking lot, Betsy asked Lance why they had to meet in such a place. "Asshole journalists following him all the time," he said. At that time, the only people following Dr. Ferrari were the Italian anti-doping police, NAS.

Later questioned under oath about the roadside meeting with his doctor/trainer, Armstrong admitted he did enter Ferrari's camper van. "I was there for a brief meeting, check body fat and body composition, and fifteen minutes later we are gone. But I understand the insinuation that I went in there and got doped up the day before Milan–San Remo. I've heard that, but that's not what happened."

The first cooling in the Betsy-Lance relationship happened a month or so later. Betsy occasionally checked out cycling websites to see what fans were saying and came across a forum where someone had said Kristin Armstrong would have a nanny when her baby was born. Betsy had talked with Kristin and knew she didn't want a nanny, so she anonymously posted a message saying this wouldn't happen. Pleased with the way she wrote the message, Betsy e-mailed it to Kristin, who got upset that someone out there thought she would have a nanny, and then Lance was mad with Betsy for showing his wife the e-mail. It was all rather silly. Frankie believed it was nothing and couldn't see there was a problem; Kristin apologized to Betsy for getting upset. The wound healed, but it left a scar.

At the start of the 1999 Amstel Gold race in Holland, Lance and Frankie got into a spiteful shouting match. It's unclear how it started but it ended with Armstrong saying he could see to it that Andreu wasn't in the Postal team for that summer's Tour de France and Andreu replying that was fine by him, as by then he would have a newborn child at home. Words spoken in anger that would, in time, be forgiven, but not forgotten. Perhaps the argument revealed more than was immediately apparent about the Lance-Frankie relationship. While Armstrong and other Postal riders, most notably Kevin Livingston and Tyler Hamilton, worked with the notorious Italian doctor Michele Ferrari, Andreu refused to. You couldn't disapprove of Ferrari, as Andreu did, and expect everything would be okay with Armstrong.

When it came to the end of his career a year later, Andreu discovered he was no different from anyone else. Suddenly he was expendable, the wrong age on a team with too many thirtysomething riders. Like most guys in this situation, Andreu thought he was worth one more year, and because Armstrong had such influence within the team, it was inevitable he would feel let down by his onetime close friend.

There was also an unpaid bonus from the 2000 Tour de France. It is traditional for the Tour winner to divide the prize money among his teammates. But because Andreu was let go by the team, he didn't get his share of the bonus given to the other riders, and though he wasn't going to harass Lance about it, he was upset at the time. Andreu returned to the Postal team as an assistant directeur sportif in 2001 and continued to write a diary for the team's website, so it was clear that there wasn't any lasting enmity between him and Armstrong.

When questions were asked about Lance Armstrong, it was natural that Andreu would be contacted by journalists. Mostly he kept his own counsel. Called by one of the *L.A. Confidentiel* authors in December 2003, Andreu said he preferred to say nothing. But the journalist was persistent and Andreu contacted Armstrong, told him what was happening, and asked what he should do. Armstrong suggested Andreu talk to the journalist, find out what he knew, and see what he was looking for. Andreu spoke to the journalist in December 2003, and though he was careful in his answers, he was surprised by how much the reporter knew. It bothered him that the journalist seemed to know about the conference room at Indiana University Hospital in October 1996, because the journalist said he'd heard Andreu and his wife, Betsy, were present when Armstrong admitted using banned performance-enhancing drugs to his doctors. Andreu refused to answer any question on that subject. Contacted by the same journalist, Betsy Andreu refused to speak on the record but privately confirmed that the hospital room confession did take place and that she and her husband heard it. Asked what she would do if ever she were subpoenaed and forced to answer under oath, she said she would not lie.

After his conversation with the journalist, Frankie Andreu reported back to Armstrong by e-mail on December 13, 2003.

> I tried to get some information for you and he finally called back. Betsy answered the phone and he started to ask her some questions but she told him she had nothing to say and then called me to the phone.

> Generally he asked some questions about Ferrari. If I ever trained with Ferrari. If there was pressure from the team to work with Ferrari. Then he asked me why not? I told him because I didn't want to. I already had a training program. He asked me if I was surprised that you were able to get fourth in the Vuelta and a couple of other stupid questions.
>
> Then he switched subjects and wanted to know about 1999 when we got in our disagreement before Amstel Gold. I told him it was nothing and it was a long time ago, he wanted to know what it was about. I told him it was something petty and small, no big deal. Then he asked if the "falling out" was the same reason that you had a "falling out" with Kevin. I told him that we still get along and I can't speak for Kevin.
>
> That was pretty much it. Give me a call tomorrow. I would have called you already it's like almost midnight for you.
>
> I don't think he will be calling back again so I don't know if I would be able to follow up at all on anything. I don't know if this helps at all but I tried. Let me know what you want me to do
>
> Talk to you later . . . frankie

After Andreu sent this e-mail, his wife took another call from the journalist, who tried to persuade her to publicly confirm that the hospital room admission took place. She again refused. He asked if she had a telephone number for Lisa Shiels, Armstrong's ex-girlfriend who was also in the hospital room. Betsy Andreu didn't know Shiels but knew that Kevin Livingston's wife, Becky, knew someone who was friends with her. She offered the journalist a deal: if he stopped ringing, she would try to get him a number for Lisa Shiels. Betsy Andreu rang Becky Livingston and told her she was being pressured by a journalist and needed Lisa Shiels's telephone number to get the reporter off her back. Betsy told Becky it would probably be better not to mention this to Kevin. Becky Livingston told her husband, who immediately contacted Armstrong.

On December 15, three days after receiving the e-mail from Andreu, Lance Armstrong replied to his former teammate.

frankie,

thanks for the help here. the more i think about this though, the harder I find it to understand why your wife did what she did. let me get this straight, you work for the team (or some form of usps), oln (where we work together on a daily basis), and all she cared about was getting left alone by this little idiot. to go around and say to becky "please don't tell Kevin" is as snaky and as conniving as it gets. i know betsy is not a fan, and that's fine, but by helping to bring me down is not going to help y'alls situation at all. there is a direct link to all of our success here and i suggest you remind her of that.

again, not to be a dick but this really stings, and i cannot for the life of me, get my arms around it.

thanks.

la

After Frankie Andreu was summoned to Lance Armstrong's hotel room at the start of the 2004 Tour de France and the meeting ended with Armstrong saying Bill Stapleton and Bart Knaggs would get in touch with him, Andreu was uneasy about where these meetings were leading. The Tour was still in Belgium and the agreed meeting with Stapleton and Knaggs took place in Charleroi after the departure of the second stage of the race. On the night before the meeting, Frankie Andreu spoke by telephone with Betsy and talked about the pressure Lance's people were applying. He was sure they were going to try to get her to make some public criticism of the authors of *L.A. Confidentiel* for suggesting she was a source. Telling her he didn't trust Stapleton and Knaggs, he said the thought occurred that maybe he should secretly tape the meeting. What did she think? She thought it was a good idea. "If it's not taped," she said, "they will deny the conversation ever took place."

Soon after the Tour de France departed Charleroi, Bill Stapleton, Bart Knaggs, and Andreu met in the sprawling parking lot close to the start. Just before meeting up, Andreu pressed the ON button of his digital recorder and slipped it into his pocket. Stapleton and Knaggs knew what they wanted from the meeting—namely, to find out how cooperative Betsy Andreu had been with the authors of *L.A. Confidentiel* and to get her to issue a statement saying she had not been a source for the book. They felt Frankie was just about on their side, but it was a question of how much control he had over his wife.

The taped conversation began with a discussion about the allegation, published in *L.A. Confidentiel,* that Lance Armstrong had admitted to using performance-enhancing drugs to doctors at Indiana University Hospital in October 1996. Stapleton says that one of the authors claimed he got that story from Betsy Andreu. "No, no, my wife is not the source. . . . LeMond's the one who gave him that story," says Frankie Andreu.

"Well, is Betsy willing to issue a statement that, or go on the record that, she didn't give . . . that she gave no comment to the guy?"

"She'll say that when he called, she said, 'No comment.' "

"The question, the question is," says Stapleton, "if she's willing to take a strong position that she . . . If she didn't give him anything about the hospital room, that's very important because it says that he's lying, that he lied about sources. . . . And if she's willing to make a statement that she'd never really testify against Lance, again, that makes him [the journalist] a liar."

"So you want her to come out with a statement saying 'I will never testify against Lance'?" Andreu asks.

"Well," says Stapleton, "I mean, she could, she could make a supportive statement about Lance. . . . She could help."

"Personally, she won't come out as saying . . . You know her and

Lance don't get along . . . and she's not gonna come out with a statement saying 'I like Lance because Lance is a good guy.' . . . She didn't, she did not tell about the hospital room. I know that for sure . . . because I never told anybody about the hospital room, you know."

A little further into the conversation, Frankie Andreu recalled the fateful day at Indiana University Hospital. "I mean, because . . . the hospital, and you know, I don't know about you, but . . . the hospital room happened, but I've never told anybody, because you know, the book, for me, what does this shit accomplish? It accomplishes nothing."

To which Stapleton replies, "Yeah."

"It doesn't do anything good for the sport," continues Andreu. "It doesn't do anything good for Lance, for cycling, or anything like that. It accomplishes nothing, because nothing's gonna happen. You know? So it's the same thing, what does it accomplish by even telling anybody, but I have never said a word about this and my wife for sure has never said a word about this."

Throughout the taped conversation, Andreu speaks about the hospital room incident as a matter of fact, and at no point does Stapleton or Knaggs challenge him about what took place. Their concern is Betsy Andreu, who they believe could be a useful witness against the book. "I know Betsy doesn't like Lance, but it's in all our interests not to blow this whole thing up," Stapleton says. Andreu agrees with this and recalls a disagreement he had with his wife when the journalist was asking questions seven months before. "I told my wife this, too, and I told Lance this, too. I went off on her and we had a whole frickin' conversation about, you know, this benefits nobody, this accomplishes nothing, we want Lance to do good because that helps me, it helps the team, it helps everything, everything just gets bigger and it helps everybody."

"And where's your mind right now?" Stapleton asked him.

"My mind right now is just to sit back and just . . . she sits . . . you know, not do anything. You know what I'm saying. She's not part of anything."

"Right."

"So, and nobody's been bothering her, and the thing is I have fucking protected Lance for a long time. Every interview I give, I frickin' talk this stuff, I say everything good, and I say I like him, you know. And then ESPN called, that's when ESPN called Betsy to go do an interview out of the blue. Flat out, she said no. Flat out, 'No, I won't do it.' "

Though Andreu repeatedly insisted his wife had not been a source for *L.A. Confidentiel,* Stapleton continued to probe. "I know that people, you know . . . people think Betsy's out to get Lance."

"Which is not true," replied Andreu. "They don't like each other, but then it stops, you know. I mean that's the extent of it."

"I hope that's right."

Soon afterward, Stapleton outlined what he hoped would happen. "Because the best result for us is to pick away at him [the author], enough that he's taken things and sort of pieced parts together and show *The Sunday Times* and show the publisher that it really is falling apart, [get an] apology, drop the fucking lawsuit and it all just goes away. Because the other option is full-out war in a French court and everybody's gonna testify and it could blow the whole sport."

"I agree," said Andreu.

Stapleton went on: "She is willing to go on the record. [She would say,] 'He called me, I talked about one thing, it was just no comment.' If that's all her statement says, it starts making him look like he's running scared, which is what he's doing. He's trying to set all these people up, to prop them up, to get courage in his mind to go forward. . . . She doesn't have to say she'll never testify against Lance. There may be an easier way to say that. So maybe we can craft something that she's comfortable with."

"That's fine," replied Andreu.

Through the conversation Andreu remained defiant, but once it ended, he felt drained and very aware that they saw his wife as a potential problem. He soon called her and told her they were on her case. She could feel his anxiety and asked him what he thought was best. "You could consider making that statement," he said. "Get that out of your head, Frankie," she said. "I'm not making any statement."

Working for OLN during that Tour de France, Andreu occasionally saw Bill Stapleton at the start or the finish of stages or sometimes near the Postal team bus. Mostly, though, Andreu tried to stay out of the way. Stapleton did call him toward the end of the race, saying he would like another meeting and that he had drawn up something that Betsy might sign. Andreu didn't call him back and it was left there, unresolved. Two months later he was contacted by Lawrence Temple from Stapleton's office and they talked. Andreu said that if they drafted a letter and sent it to him, he and Betsy would review it and decide what to do. The letter was never sent and there were no further calls. The story just went away.

In a litigation involving an insurance company, SCA Promotions, against Lance Armstrong over a year later, Bill Stapleton was asked under oath if he had spoken with Frankie Andreu during the 2004 Tour de France. "I saw Frankie at the Tour," he replied. "You know, on and off outside the bus, say hello."

The fraught seventeen-minute meeting in Charleroi had completely slipped his mind.

Chapter 17

THE STING IN THE TALE

Dog-day afternoons in Paris, sweltering heat, tourists mopping brows, televisions speaking endlessly of the drought, the locals content that in a week or so they will be gone—to the coast, to the mountains, to their places in Normandy, Brittany, the Vendée, anywhere to get away from this infernal city in August. It is late July 2005; a few days earlier summer sport in the city ended as it had done for the six previous years, with Lance Armstrong winning the Tour de France, this one his seventh. The achievement is beyond the comprehension of a people that understand the three-week race. Some celebrate Armstrong's win, others raise their eyebrows, and when Armstrong speaks on the Avenue des Champs-Elysées of those who can't believe in the miracle of his success, they shrug their shoulders. *"Si tu le dis, mon vieux."* ("If you say so, old buddy.") What they will admit is that there has been no proof of wrongdoing, no piece of evidence that you could hold up and say, "Here it is, irrefutable." Now he has won his last Tour and the sun is setting as he steers his Trek bike from the Champs-Elysées for the last time.

Those who write about cycling in Paris are restless. How can they look forward to the August break when there is so much talk of a big story about Armstrong? Something that proves he doped. Heavy stuff. They wonder who has it. *L'Equipe*? *Le Monde*? *Libération*? Each day they check one another's newspapers with a mixture of excitement and foreboding. On Tuesday, August 23, the daily sports newspaper *L'Equipe* turns the rumor into reality. The headline stretches right across the front page. It says: "Armstrong—The Lie." What is written directly beneath is even more accusatory: "*L'Equipe* has obtained the results of scientific analyses from the national anti-doping laboratory at Châtenay-Malabry, and cross-checked them against a series of official documents. Our exclusive investigation demonstrates that Lance Armstrong used r-EPO at the time of his first Tour de France victory, in 1999, contrary to what he has always said." Inside, the newspaper devotes a further five pages to their exclusive story.

At this time, Armstrong is already involved in libel suits against a French publisher, a French magazine, an English newspaper, and a small army of individuals who have, in one form or another, made accusations against him. No previous allegation, though, was as damning as this. And *L'Equipe* hasn't just called him a liar and a doper, it has also invited him to sue.

The previous day, August 22, 2005, the newspaper's editor, Claude Droussent, chaired the daily conference that began at the usual time, eleven A.M. Discussing what would be in the following day's paper, Droussent and his department heads concentrated on their coverage of the French soccer teams involved in the European Cup and a comprehensive review of Maria Sharapova's career, the Russian tennis player who had just been ranked number one in the world. As they were about to disperse, Droussent said there was a chance that late in the day something could happen to change all the plans.

"What's that?" he was asked.

"Armstrong," he said.

"Can you not tell us more?"

"A big drug story," he said. "We're waiting for final confirmation."

The various people in the room were told to say nothing, but, journalists being gossips, the story that *L'Equipe* had something big on Armstrong began to circulate. At five-thirty in the late afternoon, the journalist behind the story, Damien Ressiot, got the confirmation he was waiting for; he gave the nod to Droussent, and the *L'Equipe* news desk buzzed with the excitement generated by an exclusive breaking story. The newspaper's trust in Ressiot's investigation stemmed from documents secured by the journalist from various sources. Without these, *L'Equipe* couldn't have been so aggressive. What they had were drug test results from the retrospective examination of urine samples collected at the 1999 Tour de France. Frozen for more than five years, the samples had been reexamined in December 2004. Seventy samples collected with fifty-two still suitable for testing, they turned up twelve r-EPO positives. The scientists conducting the tests knew each sample only by its six-digit code. Without names corresponding to the codes, they had no way of knowing to whom the samples belonged. They also had no particular interest in finding out because the testing was for research purposes only.

From the start, Damien Ressiot was intrigued by the possibility that Armstrong might be one of the riders responsible for the positives. First he needed to get hold of copies of the test results produced by the French laboratory. He did that. Then he needed copies of Armstrong's official doping control forms, containing both his name and the unique six-digit number assigned to each form. Ressiot got those and by matching the doping control forms with the laboratory results, he discovered that Armstrong was responsible for six of the twelve positive results from the 1999 Tour.

Because *L'Equipe* had both sets of documents, it knew the onus would be on Armstrong to prove the documents were forged or somehow incorrect. By calling him a liar and a cheat, the newspaper went much further than any previous accuser. Sue us if you dare, the newspaper taunted. *L'Equipe* also believed the dates of Armstrong's positives made sense when set alongside the route and major battlegrounds of the 1999 Tour. The first two positives were on July 3 and 4, right at the start of the race when the leading contenders are eager to show their well-being. Armstrong won the opening 4.25-mile time trial by a staggering seven seconds. Positives three and four corresponded to the Alpine stages of July 13 and 14, and the final two came on July 16 and 18, as the Tour entered the Pyrenees.

Among the many reactions to the story, perhaps the most frequently expressed reservation concerned the timing. Was it a coincidence that *L'Equipe* ran such a damaging story four weeks *after* Armstrong won his seventh and final Tour de France? The newspaper and the Tour de France are owned by the same parent company, the Amaury Group, and if the story had been printed before or during the race, the Tour would have been damaged. Ressiot vigorously denies the timing was commercially driven and insists the intention had been to run the story during the Tour de France but he didn't get the confirmation he needed until three weeks after the race ended.

Damien Ressiot had been a writer with the respected soccer magazine, *France Football,* from 1990 to '97. He then spent two years working on soccer for *L'Equipe* before being given responsibility for the newspaper's coverage of doping. With the change in remit, Ressiot was expected to come up with investigations of his own, and over a number of years, he developed good contacts inside the world of doping and drug testing. In January 2005 he heard the French anti-doping laboratory was involved in research to improve its urine test for r-EPO. A few days later he learned scientists at the laboratory had retested samples from the 1998 and 1999 Tour de France. They chose those two Tours because they believed that at the time, many cyclists were using the then undetectable r-EPO. Their assumption was correct and as soon as Ressiot learned that numerous samples were positive for r-EPO, he began to pay attention.

His interest was triggered by the possibility of a story that could involve the perennial Tour winner, Armstrong. Ressiot was no cycling specialist, but he was French and knew well that Armstrong won his first Tour in '99, and as race leader for almost two weeks, he would have been tested a lot. So if a number of the '99 samples showed evidence of r-EPO, as Ressiot was being told, there was a good chance that one or more of Armstrong's samples was among them. The journalist was motivated by the desire to break an important story but he also wanted to challenge the law of silence that informed cycling's attitude toward doping. In his dealings with professional cyclists, he found them unresponsive to inquiries about doping and he believed they felt entitled to act outside the rules of their sport. As for Armstrong, Ressiot's opinion of the guy had been colored by an incident between the champion and a small-time Italian rider, Filippo Simeoni, on the third-to-last stage of the 2004 Tour de France.

When Simeoni surged in pursuit of six breakaways soon after the start at Annemasse that Friday, it should have passed unnoticed. The Italian was no threat. Armstrong saw it differently and chased him down. This was unusual because the wearer of the yellow jersey does not normally leave the shelter of the peloton to follow a modest équipier. After Simeoni and then Armstrong joined the six breakaways, the race leader told the group that they would not be allowed to escape if Simeoni remained among them. From Armstrong's point of view, the Italian was the rat who testified against Michele Ferrari and a key witness in the case against the doctor. That made him an enemy. Two of the six breakaway riders suggested to the Italian he should drop back and, not wanting to destroy the chances of the six, Simeoni did as he was told.

He had appeared as a witness in the Ferrari trial on February 12, 2002. "From November 1996 to November 1997, I was treated by Ferrari," he told the court. "Even before that, I had taken doping products. Ferrari gave me a work schedule of increasing toughness; r-EPO was spoken about from the first moment. That year, I was effectively taking r-EPO on his instructions. Later, in March and April,

Andriol was spoken of. I needed to take it after intense training sessions. Ferrari also told me to be careful about taking testosterone too close to races due to the risk of being tested positive. I have never been tested positive. To avoid anti-doping problems, Ferrari told me to use Emagel on the mornings of tests and to use another product the night before to lower my hematocrit level."

In an interview with the French daily newspaper *Le Monde,* Armstrong called Simeoni a liar. Simeoni sued for libel and a year or so later, the case was settled out of court with both parties agreeing to a confidentiality clause preventing them from discussing the terms of the settlement. It is believed Armstrong paid Simeoni 100,000 euro ($125,000). Why should Armstrong have felt compelled to pursue Simeoni at an inconsequential moment in the Tour de France and draw upon himself widespread criticism for an act perceived to be mean-spirited and bullying? Speaking in 2003, Cédric Vasseur, a former teammate of Armstrong's at the U.S. Postal Service team, offered a portrait of the champion that offers some clues.

"More than anything else, Armstrong is very smart. He allows very few people to get too close. He leads a separate life; he changes masseurs every one or two years to protect himself. Since 1998, Emma O'Reilly, Freddy Viaene, and Ryszard Kielpinski followed one another. He organizes his schedule so no one can guess where he will be. He's an intelligent guy. He knows he is being watched. Within the team, he likes to be considered the boss, to be taken for God. The only time he is pleasant is in January when he's not on the bike. He doesn't tell you things directly; he'll go through somebody else—in most cases Bruyneel. That's the middleman. For example, in the Tour de Suisse that he won, he found himself without teammates in the final part of the first stage. The next morning we had a terrible meeting. Bruyneel gave us a real going-over. Armstrong was sitting there like us, not saying anything. But everything Bruyneel was saying had come from Armstrong.

"He is also a perfectionist. He leaves nothing to chance. He is also paranoid, particularly so since the September 11 attacks. He's got a bodyguard to make sure a bin Laden fan doesn't slit his throat.

But maybe the biggest thing is his ability to intimidate. In cycling, if you want to win races, you have to crush the others. Armstrong taught me to be nasty. He runs on aggression, anger, intimidation, and the fear he can strike in others. He doesn't kick up a big fuss but he goes into a state of cold, suppressed anger. He wants to impose the law of the strongest, and most of the time it works."

Damien Ressiot was among those at once intrigued and appalled by Armstrong's desire to punish Simeoni. "The case had a profound effect on me," he said. And so he focused his efforts on finding the identities of the twelve positives from the 1999 Tour de France. "At first, the only thing I had were the lab results but with no names. That was just one half of the jigsaw and the other half was the harder one to find. I went looking in different places and always with the greatest discretion. If I had approached the wrong person, those documents might have been burnt." There were four possible sources for the doping control forms that Ressiot needed. Copies are sent to the UCI, the French Ministry of Youth and Sport, and the French Cycling Federation. Finally, the doctor who takes the sample is permitted but not obliged to keep a copy of the doping form.

Without a source prepared to hand over the documents, Ressiot came up with a plan to get the UCI to volunteer the relevant information. He contacted cycling's governing body, said he was doing a story on Armstrong, and suggested that if he could see the rider's doping control forms, it could help to show Armstrong had ridden clean. Ressiot's approach was clever, if disingenuous. He gave the impression his story concerned the use of Therapeutic Use Exemptions (TUEs), which underpin an accepted and legalized form of doping in professional sport. For example, in the 2006 Tour de France, 60 percent of the competitors had TUEs allowing them to use a banned product. Armstrong had always said he did not have a TUE for any product, and if this were true, Ressiot said he was ready to write it. The truth could be established by showing the doping control forms to the journalist. Ressiot spoke with a number of people at the UCI and liaised also with Donald Manasse, one of Armstrong's French lawyers.

His sales pitch focused on the benefit to Armstrong of a story showing he did not use medical exemptions. Ressiot contacted Hein Verbruggen, the then president of UCI, who had little time for the journalist but saw merit in a story that would portray Armstrong in a favorable light. During a phone call that took place when he was in China on IOC business, Verbruggen told Ressiot he was in favor of showing him Armstrong's doping forms but the decision would ultimately be the rider's. Initially, the Texan was reluctant and his directeur sportif, Johan Bruyneel, was dead set against giving Ressiot anything. They knew his investigation was not as innocent as he made out but mistakenly guessed his search was for a therapeutic exemption allowing Armstrong to use a banned product. He had no such dispensation, and as the forms proved this, Verbruggen and eventually Armstrong were persuaded that disclosure could only help the rider. They agreed to show Ressiot the forms. The journalist also sought access to Armstrong's medical records, a request that tallied with the idea he was looking for something connected to a TUE. That request was turned down.

Ressiot traveled to the UCI headquarters at Aigle, Switzerland, in July 2005. Because Verbruggen and Armstrong approved, doors were opened for the journalist and documents handed over. In fact the UCI medical director, Mario Zorzoli, could not have been more helpful and gave Ressiot Armstrong's doping control forms from the 1999 Tour de France and every subsequent Tour up to 2004. With a specific agenda, the journalist concentrated on the '99 race and after a visit that lasted two hours, he left with the documents he sought. Two elements of a three-part story had been covered; Ressiot knew of the twelve positives from the retesting of the 1999 samples, and second, he had Armstrong's doping forms for '99. All that remained were official copies of results from the lab, which Ressiot knew had to have been passed on to the French sports ministry and the World Anti-Doping Agency.

There was no urgency about the lab's research work, and through the first six months of 2005, the results from the retrospective tests were not collated. Toward the end of July, Ressiot learned the results were being compiled, and, suspecting they would be sent to the French sports ministry and WADA, the journalist canceled his summer vacation to the Ile d'Oléron off the west coast of France. At five-thirty P.M. on August 22, Ressiot received what he calls "the final confirmation" of what would be a sensational story about Lance Armstrong, r-EPO, and the 1999 Tour de France. While he refuses to be specific about what he knew before receiving the official set of results from the laboratory tests, Ressiot admits the results provided confirmation of what he already knew. Comparing the identification numbers of the positive tests from the lab with Armstrong's official doping control forms, he found six correlations. Much of his story had already been written, and with the second set of documents (the official test results from the laboratory), *L'Equipe* made the decision to print and be damned.

Ressiot was accused of having been spoon-fed the documents by virtue of his newspaper's relationship with the Tour de France and he was also criticized for attempting to make a laughingstock of Hein Verbruggen. From the official world of cycling, there were two congratulatory messages—one from Daniel Baal, the former president of the French federation, and another from Armand Megret, the French federation's official doctor. The journalist himself was left with mixed feelings about Armstrong. "He's no saint, but I totally understand why he did it. How could anyone have competed in the 1999 Tour de France and not engage in a practice allowed by the whole system? Can all of his successes be explained by doping? I don't know. What I can't stand is the deceit." Ressiot was unequivocal about the part played by sports journalists. "We sell stories of extraordinary achievement but when we learn they are not that, we don't like to take them back. I feel I have done my job as a journalist, and as a profession, we need to stop building dreams on false premises."

L'Equipe's story generated wide-ranging debate. Was this the evidence that proved Armstrong had doped? Having long been Armstrong's defender, the then Tour de France director Jean-Marie

Leblanc was persuaded by Ressiot's investigation. "For the first time," said Leblanc, "and here we are no longer speaking of rumors or insinuations, these are proven scientific facts; someone has shown me that in 1999 Armstrong had a banned substance called r-EPO in his body. He owes explanations to us, to everyone who followed the Tour. What *L'Equipe* revealed shows me that I was fooled and we were all fooled."

Asked whether the samples identified as his were actually his, Armstrong said, "I have no idea. I can only believe that they either are not mine or have been manipulated." The first thesis, that the samples were not in fact Armstrong's, is virtually insupportable. When scientists at the laboratory conducted their research testing, they dealt with samples from the 1998 and '99 Tour de France that were each identified only by a six-digit number. At no point did anyone in the laboratory have access to the doping forms that would have allowed them to match numbers to names. The documentation allowing Ressiot to match Armstrong's name with the laboratory numbers came from the UCI headquarters in Switzerland.

Confronted by the evidence in the *L'Equipe* report, Armstrong concluded it was more likely the samples were spiked than the doping forms were forged. In his interview on *Larry King Live,* he was asked if he had kept his copies of the doping control forms from the '99 Tour. "You get a carbon copy of the test results or the test form that day. I wouldn't have them from 1999. But that's not what was manipulated. What was manipulated was the urine. What was put in the urine? Who was there when . . . I don't think the papers were manipulated." Manipulation is a serious allegation against the laboratory, and it raises the question of what motive the lab would have had to spike Armstrong's samples and also the question of whether it could have been done, given they did not know which samples were his. However, it is possible to add r-EPO to a urine sample to produce a false positive.

On this occasion, as on many others, Armstrong referred to a witch hunt against him, and suggested the anti–United States feeling, prevalent in France at the time, was the reason for the difficulties he faced. It is possible, however, to work out the chance of the French lab's being able to successfully spike his urine. The starting point is that the laboratory staff did not have names to go with their numbered samples, but since each sample is dated, someone at the lab could have figured out the days upon which Armstrong was tested. There would still have been the difficulty of spiking the correct sample because Armstrong's would have been one in a batch of four, five, or six, depending upon the particular day.

On the first occasion when Armstrong was positive, all four samples collected that day contained r-EPO. Therefore, if the intention was to spike his sample, the chance of doing so successfully was four out of four. On the second day, five samples were analyzed but only one contained r-EPO, which means the chance of the lab correctly spiking Armstrong's sample was one out of five. On the third occasion, six samples were analyzed but only Armstrong's was positive: the chance of the lab spiking the correct sample was one out of six. On the fourth day, four samples were tested and two contained r-EPO. That gave the lab a two out of four chance of manipulating Armstrong's sample. For his fifth positive, Armstrong's sample was the only one of four to contain r-EPO. On the sixth day, Armstrong's sample was one of two positives from the four analyzed. The chance of the laboratory successfully spiking Armstrong's sample on that day was two out of four.

There is nothing complex about the mathematical equation that calculates the likelihood of the laboratory manipulating Armstrong's urine samples. The likelihood can be estimated by multiplying the possibility of the lab successfully spiking Armstrong's sample for each of the six positives: $4/4 \times 1/5 \times 1/6 \times 2/4 \times 1/4 \times 2/4 = 16/7680$ $= 1$ in 480.

In an interview with Larry King and Bob Costas, aired on CNN's *Larry King Live* two days after the story appeared, Armstrong was asked if he intended to sue the French newspaper.

Costas: Do you plan legal action?

Armstrong: That's the most commonly asked question in the last three or four days, and it's a possibility. We have to decide who we're going to pursue, whether it was the lab, whether it was *L'Equipe,* whether it was the sports minister, whether it was WADA. All of these people violated a serious code of ethics.

King: Why not sue them all?

Armstrong: You know, lawsuits are two things. They're very costly and they're very time-consuming. And fortunately, cycling has been great to me and I have the money and the resources to do something like that, but you know, I'm retired.

Costas: You've done it before. You have civil cases pending.

Armstrong: I do. Absolutely.

Costas: You've been litigious before when you felt it was justified.

Armstrong: Yes and you know what, at the end of the day when you sue somebody, it just keeps a bad story alive forever. It gives them the opportunity to say, "Oh, we found this. Oh, we found that." It gives them more credit than they deserve.

Perhaps Armstrong was tired of what must have seemed like eternal litigation. At the time, he was still involved in actions against *The Sunday Times,* the publishers of *L.A. Confidentiel,* Stephen Swart, Emma O'Reilly, and his former personal assistant Mike Anderson, and he was also embroiled in a particularly time-consuming litigation against the Dallas-based company SCA Promotions. Whatever factor or combination of factors influenced him, Armstrong decided against suing *L'Equipe,* and all the other cases would be settled or dropped before a judge was allowed to deliver a verdict.

The reaction of the UCI to the *L'Equipe* story was one of indignation, even outrage. What troubled cycling's governing body was not the allegation that Armstrong used banned drugs but the fact that research results were leaked and formed the basis of a story that called Lance Armstrong a cheat. The reaction of the World Anti-Doping Agency was very different, and two days after the story appeared, it wrote to the UCI suggesting they hold an inquiry. Written by WADA's chief executive, David Howman, the letter emphasized that the issues raised by the story were the UCI's responsibility. "As these matters precede WADA, and of course the WADA Code, jurisdiction rests with you as a responsible anti-doping organisation. Can we ask, please, what steps you intend to take? We are at your disposal for any assistance you may seek, and are happy to work with you accordingly."

Four days later the UCI issued a press release stating it would assess the situation, "whilst regretting, once more, the breach of confidentiality principle which led to the divulgence of this information." The statement went on to say the UCI would deliver its conclusions regarding the *L'Equipe* story within ten days. The then president of the UCI, Hein Verbruggen, wrote to WADA on August 30, one week after the publication of the Ressiot story, and categorically said nothing would be done. "As you can expect from us, we will not take any action based upon a press article and most definitely not upon articles from Mr. Ressiot, of which we know his attitude towards cycling and the UCI." In a second letter by Verbruggen, written on the same day, and again sent to Howman, the head of UCI reiterated his reluctance to authorize an investigation.

> You ask us to investigate the matter on the basis of a newspaper article.
>
> As far as I understand, the analyses that are referred to were made at the request of WADA for research purposes. The laboratory confirmed in a press statement that the research results were given to you anonymously and could not be used for disciplinary purposes.

David, in a WADA-initiated research program conducted in a WADA-accredited laboratory, the most essential standards of confidentiality have been disregarded.

Confidential information of this study became available to the press.

And now you ask me to investigate . . . ???

The letters traveled to and fro. UCI sent WADA a list of questions framed to ascertain how Ressiot had gotten his information, and WADA in turn expressed its concern about the UCI's lack of interest in the substance of the *L'Equipe* story. Did the governing body care if Armstrong had or had not used r-EPO? On September 14, WADA chairman Dick Pound wrote to Hein Verbruggen expressing his lack of faith in the UCI's approach.

The questions you have directed at WADA have been generally accusatory in nature and have been surrounded by several statements and assertions with which WADA is unwilling to be associated. Every question points in one direction only, namely how the various elements of the *L'Equipe* story were obtained by the reporter. Not a single one focuses on the issue whether or not the allegations are true or whether they are false. . . . All of your investigative efforts, based on what we have seen, appear to be directed at finding someone to blame for the disclosure of information that you seem to regard as confidential and the statements attributed to you in the media are to the same effect.

Two days later Pound told reporters he believed the key doping control forms, allowing Ressiot to match names to the leaked research results, were given to the journalist by the UCI. In response to this, the UCI admitted it gave Ressiot one of Armstrong's doping forms from 1999 but not the others. They later admitted they did in fact supply Ressiot with all fifteen of Armstrong's doping forms from the '99 Tour and suspended its medical chief, Dr. Mario Zorzoli.

On September 20, the Association of Summer Olympic International Federations and the IOC Athletes' Commission wrote a joint letter to WADA complaining about irregularities perpetrated by both the French anti-doping laboratory and the World Anti-Doping Agency in the controversy surrounding Armstrong's samples, and went so far as to ask WADA to suspend the accreditation of the French lab.

Over a month had passed since the *L'Equipe* story and still the controversy raged. The UCI's starting point was that an investigation was unnecessary, but as the weeks passed, that position became untenable. On October 6, it announced that the Dutch lawyer Emile Vrijman would undertake an independent inquiry into the issues raised by Ressiot's story. Vrijman is an attorney at Scholten c.s. advocaten in The Hague and had served for ten years as codirector of the Netherlands Centre for Doping Affairs (NeCeDo). In an interview published in the Danish newspaper *Politiken* on July 9, 2006, Vrijman helped explain how he came to be appointed as the UCI's independent investigator. "Yes," he said, "I am a very good acquaintance of Hein Verbruggen's, because of my former job as director of the Dutch anti-doping agency."

On November 9, 2005, the UCI issued a "Letter of Authority" specifying Emile Vrijman's mandate and outlining the conditions under which he would work. By then the investigator had already begun and in late May 2006, his 132-page report was severely critical of both WADA and the French laboratory for their roles in the controversy. Armstrong claimed the report completely exonerated him but Vrijman, in his *Politiken* interview, specifically stated the report did not exonerate Armstrong, because determining his innocence or guilt was not part of the investigator's mandate. The report concluded there could be no disciplinary action taken against the athletes responsible for the positive tests because the researchers at the French laboratory had not adhered to the proper standards for drug testing. That was something Jacques de Ceaurriz and his staff at the Châtenay-Malabry laboratory understood from the beginning;

they tested for research purposes and were not obliged to adhere to protocols necessary for official drug testing. Vrijman devoted countless pages of his report to the many instances where the laboratory did not adhere to strict drug testing regulations.

The report also questioned the laboratory's right to use old urine samples for research purposes. "The laboratory representatives stated they had never considered whether or not they were actually allowed to use these urine samples for research purposes and consequently had neither asked the riders nor the UCI for any permission for their use, nor clarified ownership [of the samples]." Another important point for Vrijman was why the World Anti-Doping Agency wanted the French laboratory to include the code numbers of the samples when the research results were being sent to them. Implicit in Vrijman's emphasis on this point is the insinuation that WADA had an anti-Armstrong agenda. In reply, the world anti-doping body explained its request for code numbers as follows: "WADA can not imagine that the UCI would not have wanted to preserve the possibility of a longitudinal study analysis of the abuse of EPO and would not have wanted to know who was abusing EPO at that time among its riders."

Vrijman invoked the Declaration of Helsinki, which governs medical research involving human subjects, and the Oviedo Convention, which protects human rights, to further his case that the French laboratory and WADA behaved improperly. From his reading of the IOC medical code, Vrijman surmises that as a laboratory is bound to store all samples for a minimum of two years and positive samples for five years, they should relinquish ownership after that point, unless the relevant governing body (in this case, the UCI) requested it to do otherwise. The 1998 and 1999 samples were more than five years old when the laboratory decided to retest them, and according to Vrijman, the lab may not have been entitled to have them in storage at that point.

Vrijman also criticized the lab director, Jacques de Ceaurriz, and Dick Pound for public comments they had made in the aftermath of the *L'Equipe* story. According to the investigator, they should not have confirmed that six of the positive results related to Lance Armstrong. In a comment made to a journalist from the Dutch daily newspaper *de Volkskrant,* De Ceaurriz offered an alternative viewpoint: "Important information," he said, "should not be allowed to be buried because of medical ethics, which do not apply to athletes anyway. They are not patients. The pretence of protecting the athlete protects especially those who cheat. The new Code should protect athletes who do not cheat."

Though De Ceaurriz insisted the research testing could not be the basis for bringing sanctions against those riders responsible for the positives, he was adamant the results themselves were correct. He made this clear in interviews with *L'Equipe* and *de Volkskrant.* "There is no possible doubt about the validity of the results, even though the analysis was carried out five years after the samples were taken," he told *L'Equipe.* To the journalist from *de Volkskrant,* he said, "The test results are what they are. By coincidence they happen to belong to the winner of the 1999 Tour de France. They could also have belonged to someone who did not win the Tour."

Emile Vrijman's report was intended to end the controversy arising from the *L'Equipe* investigation, but it failed to do that. Ironically, a report that criticized in the strongest terms the leaking of confidential information to the media was itself leaked to the press. It was reported in Dutch newspapers before the UCI, which was paying Vrijman, received its copy. "The UCI firmly deplores the behaviour of Mr. Vrijman" was the reaction of cycling's governing body to the man it had commissioned to conduct its inquiry. And WADA was seriously unimpressed by the quality of the investigation. "The Vrijman report is so lacking in professionalism and objectivity that it borders on farcical," said Pound. "Were the matter not so serious and the allegations it contains so irresponsible, we would be inclined to give it the complete lack of attention it deserves." In a subsequent statement, WADA was scathing about the report. It recalled a meeting between Dick Pound and Hein Verbruggen at the Turin Winter

Olympics in February 2006, during which Verbruggen advised Pound he had seen a draft copy of Vrijman's report. "The report," he told Pound, "will make extremely bad reading for the World Anti-Doping Agency." In response Pound advised Verbruggen that WADA had not yet been approached by Mr. Vrijman with any requests for information or for interviews.

WADA's statement then referred to the process used by the French lab in its research work, which was not the process used for analyzing samples for the purpose of doping control. The failure to make this distinction, according to WADA, was the central flaw of Vrijman's report and had led to "ill-informed and incorrect outcomes." In its statement, WADA posed the following question: "Mr. Vrijman does not inquire at all into why Mr. Armstrong gave his consent, through his advisers, to UCI to provide 15 doping control forms to the *L'Equipe* reporter who was the author of the article published on 23 August. Mr. Vrijman does not likewise ask or inquire in any depth of UCI management and executives of why they sought Mr. Armstrong's consent, and why they authorized the release of the documents. . . . That failure indicates both a lack of professionalism and a distinct lack of impartiality in conducting a full review of all the facts. Indeed, despite Mr. Verbruggen's concession that all 15 forms came from UCI, Mr. Vrijman only suggests it may have been more than one. Why did he fail to review all the files, and interview the responsible personnel?"

WADA concluded its assessment with a dismissal of the report's author. "When the process is so flawed as it is to date, there can no longer be professional confidence in the author."

This wasn't Vrijman's first experience of an anti-doping controversy. Fourteen years before, during his term as codirector of the Dutch anti-doping agency, he was at the center of a dispute involving the world-class German sprinter Katrin Krabbe. In January 1992, Krabbe, accompanied by two German teammates, Grit Breuer and

Silke Moller, trained in South Africa. While there, they were subjected to an out-of-competition drug test and all three produced urine samples that were biologically identical, indicating they had come from the same person. The samples were collected on January 24 and a week later the German athletics federation suspended the athletes and their coach Thomas Springstein. Krabbe pleaded her innocence and with her lawyer, Reinhard Rauball, claimed someone must have interfered with the dope testing kits containing the urine. "Impossible," said the late Manfred Donike, head of the Cologne laboratory that had analyzed the samples.

Rauball then attempted to prove the samples could have been manipulated. He produced other plastic dope testing kits from the same manufacturer, the English company Envopak, that contained precisely the same codes as the ones used to collect the samples in South Africa. Krabbe's lawyer got these kits from Emile Vrijman, who used his position as director of the Dutch anti-doping agency to get Envopak to supply them. Interviewed by Lars Jorgensen, a journalist with the Danish newspaper *Politiken,* Vrijman did not deny his role in the Krabbe doping case. Asked why he helped an athlete accused of cheating, Vrijman explained, "Because her Dutch agent, Jos Hermans, asked me to and because I wanted to show that the credibility of the international anti-doping movement was under threat as long as it was possible to manipulate the plastic doping kits." As director of NeCeDo, it was a loophole he could expose. "I just wrote to the company and ordered them. I never tried to hide anything, and by doing this, I proved that outsiders could easily manipulate urine samples by coming up with identical kits."

Not everyone took as sanguine a view of Vrijman's action. "I liked Emile," says Mark Gay, the International Amateur Athletics Federation (IAAF) lawyer on the Krabbe case. "He was good fun and enjoyable company if you were having a drink in the bar after a meeting or whatever. But I was surprised by what he did in the Krabbe case." Bryan Wooten was doping officer for the IAAF at the time. "It caused a lot of resentment amongst anti-doping people at European

Council level that Vrijman should have done this," says Wooten. "It wasn't right."

Rens van Kleij, a former colleague of Vrijman's who still works at NeCeDo, told the journalist Jorgensen, "Emile Vrijman abused his position as anti-doping director of NeCeDo and helped Katrin Krabbe to prove that people from outside could have interfered with her urine test. I don't know why Emile Vrijman helped Krabbe but he has often been criticised for helping athletes accused of doping. Since 1998 he has been asking questions about all of the procedures used in doping cases."

The chairman of WADA's Health, Medical and Research Committee, Professor Arne Ljungqvist, was head of the IAAF's medical commission from 1981 to 2003 and was involved in the Krabbe case. Fourteen years on, he spoke about his feelings at the time. "We were very upset about it. We knew someone from NeCeDo was involved but we never managed to be sure who. I didn't know it was Emile Vrijman. What he did was disgusting and unethical and it is very interesting in the light of his report in the Lance Armstrong case."

One month after escaping punishment for the South Africa tests, Krabbe and Crit Breuer tested positive for the anabolic steroid clenbuterol. At the time, clenbuterol was regarded as a veterinary product and not on the IOC's list of banned substances. However, Krabbe and Breuer were given three-year bans for "unethical behavior," a punishment that led to a protracted lawsuit taken by Krabbe against her national federation. In 2001, nine years after the clenbuterol discovery, the German federation was ordered to pay Krabbe 615,000 euro ($520,000) in compensation for earnings lost because of her ban.

For his part, Emile Vrijman doesn't see that he did much wrong in the Krabbe case. "I know there were many people who were happy that I showed the system, which we were paying a lot of money for, was not protected from abuse. I guess the Dutch sports minister would not have allowed me to stay on as director of NeCeDo for almost ten years if I had done anything improper."

On June 9, ten days after the release of the Vrijman report, Lance Armstrong wrote a letter to the president of the IOC, Jacques Rogge, and copied it to each member of the IOC's executive committee. The letter ran eight pages and expressed Armstrong's sense of vindication following the publication of the Vrijman report as well as his disgust at the behavior of the World Anti-Doping Agency and the French anti-doping laboratory at Châtenay-Malabry. Most of all, the letter expressed Armstrong's contempt for Dick Pound, and his heartfelt plea that Rogge and the IOC executive committee remove the Canadian from his position as head of WADA.

About Pound, Armstrong wrote: "The misconduct of the laboratory apparently all occurred at the direction and insistence of Dick Pound. While that does not excuse the laboratory, it evidences a completely unacceptable situation in which laboratory officials are willing to break rules and break laws rather than resist demands by WADA. The misconduct of Dick Pound all came as no surprise to anyone in the Olympic movement." A little further on, Armstrong is no less restrained. "Dick Pound is a recidivist violator of ethical standards. His conduct in this matter is completely reprehensible and indefensible." Such was Armstrong's sense of being wronged that he informed Rogge, "I have been advised that I can pursue legal action against WADA, the French Ministry, the French laboratory, and their officials."

As with the *L'Equipe* story, there would be no litigation.

Before calling on the IOC to sanction Pound, Armstrong revealed what motivated him. "Despite all the damage that I have endured, I have no interest in allowing further damage to be caused to the Olympic movement in general or the credibility of international drug testing in particular." The Olympic movement, he felt, would be protected by the removal of Pound. "If the individuals responsible do not accept responsibility and yield their positions voluntarily, those individuals must be suspended or expelled from the Olympic movement." The final sentence of the eight-page letter thanked

Rogge and the executive committee "for your immediate attention to this matter."

In reply, IOC president Jacques Rogge said he would welcome "an independent investigation" into the allegations contained in the *L'Equipe* story. The IOC boss had given his opinion on the Vrijman report.

Chapter 18

THE HOSPITAL ROOM—PART TWO

In his autobiography, *At the Table,* Bob Hamman tells a story about a controversy in 1975 that offered an insight into the passion he brought to his sport. Back then Hamman was the best bridge player in the world, a distinction he has enjoyed for much of the last three decades. Eleven times world champion, he was once described as the Babe Ruth of his game. His improbable story from 1975 concerned cheating. That year, bridge's world championships were held in Bermuda, and the tournament was played against a backdrop of intense hostility. Before it began, America's nonplaying captain Freddy Sheinwold wrote an article published in the magazine *Popular Bridge,* in which he accused two members of the then world champion Italian team, Gianfranco Facchini and Sergio Zucchelli, of cheating. Shock waves tumbled across the sport's normally sedate surface.

Within the bridge community, others had heard rumors about the Italians but it was convenient to pretend they weren't true. After all, cheating at bridge is not unlike shoplifting: middle-class people prefer not to speak about it. Especially on the old continent, where Sheinwold was pilloried for making unsubstantiated allegations, and where the Italian bridge federation called for his removal from the U.S. team. Hamman, a U.S. team member, fully supported Sheinwold. The fuss eventually died down and nothing much might have come of Sheinwold's finger-pointing if it wasn't for the presence in Bermuda of Bruce Keidan. A modest player from Philadelphia, Keidan had convinced *The Philadelphia Inquirer,* for whom he wrote about bridge, to send him to the world championships.

In Bermuda, Keidan volunteered for duty as a monitor recording the bids of various matches. Always needing such help, the organizers accepted Keidan's offer and while observing the Italy-versus-France match, he noticed strange goings-on between Facchini and Zucchelli. Facchini sat with his feet wrapped around the legs of his chair, but for each auction he would stretch one of his legs forward and touch one of Zucchelli's feet; right foot to right foot, right to left, left to right, left to left, in what seemed a series of prearranged signals. Zucchelli never once pulled his feet away, which might have been the natural reaction to accidental contact.

Keidan had been aware of the suspicions about the Italians but he had also been skeptical. Convinced they were now true, he reported what he had seen. Others were assigned to monitor Facchini and Zucchelli and witnessed the same illegal collusion. Disgusted, Hamman believed the Italians should and would be thrown out. But the World Bridge Federation Appeals Committee merely reprimanded the culprits for "improper conduct" and allowed them to carry on. The authorities did put coffee tables under the card tables to make it impossible for a player to touch the foot of his partner. As bad luck would have it, the United States lost to the Italians in the final, and the memory of what followed is embedded in Bob Hamman's soul. "Our hearts were in our throats. We knew it was doom for our team. And the room was cheering for Italy. Americans were cheering for Italy. Bobby [his playing partner Bobby Wolff] started crying. He was sick to see our compatriots rooting for the cheaters. It was more than he could bear. I was momentarily, perhaps morbidly, enthralled by the scene and I forgot where I was. When I turned round, Bobby was gone.

"Sitting through the victory banquet—watching those cheaters claim a world championship—was one of the most difficult things I've ever done in bridge. I've lost a lot of championships through the years, but defeat was never so bitter for me as on that occasion. We made our way to the victory banquet, all suited up in tuxedos but feeling anything but jaunty. We were bitter, but we were also determined to show class in the face of this outrage. We weren't going to be poor losers on top of everything else."

For Hamman, losing was one thing. Being cheated out of something was quite another.

Eleven years after Bermuda, Hamman set up SCA Promotions, a business venture that played to his gift for quickly calculating the odds. Fundamentally, SCA is a bookmaking business: a legal, respectable, profitable firm that runs a business not much different from the guy who covers your bet at the racetrack. Hamman's company underwrites the risk taken on by organizations for various one-off payments; for example, a golf club manufacturer sponsors a talented rookie and agrees to a clause in his contract to pay him five million dollars should he win a major championship in his first or second season. The sponsor knows this is unlikely but doesn't wish to carry the risk. So they strike a deal with SCA Promotions, paying the company an agreed premium and passing on the risk.

SCA is the world leader in its corner of the risk market, having guaranteed $12 billion in prizes and bonuses and paid out more than $134 million over the last twenty years. In December 2000, Kelly Price, an insurance broker with the company Entertainment and Sports Insurance Experts, contacted SCA's Chris Hamman. As well as being Bob Hamman's son, Chris Hamman is an underwriting expert at the company. He had done business with Price before and so he listened. The proposed deal involved the cyclist Lance Armstrong and the possibility of SCA taking on some of the risk carried by Tailwind Sports, owner of the U.S. Postal Service cycling team, should Armstrong go on winning the Tour de France. At that point, he was a two-time winner. SCA was being asked to quote on a deal that would give Armstrong $1.5 million for winning in 2001 and 2002, a further $3 million for winning in 2003, and a final $5 million bonus payment if he maintained the winning streak and won a sixth consecutive Tour in 2004. Knowing little about professional cycling, Chris Hamman felt SCA shouldn't get involved.

If the company could turn back the clock, negotiations would end at that point, Bob Hamman would go with his son's instinct and tell Price he is not interested. But the elder Hamman had become a successful businessman because he'd found ways to make deals and didn't often fold his cards. What did it matter that this sporting event was taking place in faraway France? The questions didn't change. How many riders would start? How many of them had a chance of winning? How many guys had previously done what Armstrong was attempting? What price were bookmakers quoting on him doing it? You worked out the figures, you tendered for the job, and so it was that Bob Hamman, one of the shrewdest players in the game of sporting risk, overruled his son and made one of the worst decisions of his business life. It is what Hamman self-deprecatingly calls "the dumbass factor" and it comes into play one deal in every hundred, he says. The contract, when it was drawn up in early January 2001, stipulated that for a premium of $420,000, SCA would pay Armstrong staggered payments amounting to $9.5 million should he win every Tour de France from 2001 to 2004. When formalized, the agreement became known as Contingent Prize Contract 31122.

Foolishness is no reason for not honoring one's debts, and for the first three years of the four-year deal with Tailwind, SCA wrote the checks due to Lance Armstrong. In 2004, *L.A. Confidentiel* was published in France and the mood changed at SCA. John Bandy, one of SCA's in-house lawyers and a fluent French speaker, bought a copy of the book and translated the most relevant sections for company president Hamman. After reading the book, Bandy felt the allegations made in *L.A. Confidentiel* warranted serious investigation. Hamman went through Bandy's translated passages, amounting to almost one hundred pages, and agreed SCA had every right to check out the claims against Armstrong before paying over the final five-million-dollar bonus payment to the Tour champion. When SCA did not immediately write the five-million-dollar check, litigation was inevitable.

In the weeks after Armstrong's victory in the 2004 race, Hamann went to work. His first play was to lodge five million dollars in a custodial account with JPMorgan to show his company had the financial wherewithal to pay Armstrong. Perhaps surprisingly for the president of a major international company, Hamman took it upon himself to make many of the inquiries concerning the claims against Armstrong. First he traveled to Europe to meet the authors of the book. He then journeyed on to Auckland in New Zealand, where he tracked down Stephen Swart, an important source for *L.A. Confidentiel.* Soon he made telephone contact with Greg and Kathy LeMond in Minnesota, and then with the Andreus in Michigan. He later spoke with Emma O'Reilly in England and let it be known he was interested in talking with anyone who had something to tell about professional cycling, the U.S. Postal Service team, or Lance Armstrong. From every source quoted in *L.A. Confidentiel,* Hamman heard the same story: they stood by what they said.

Of all the accusations in the book, the alleged hospital room confession was the one that intrigued Bob Hamman. If, as the authors strongly suggested, Armstrong admitted doping to his doctors at Indiana University Hospital in October 1996, that meant he had cheated. Hamman's view was that if SCA had known about the hospital room allegation, it never would have done business with Tailwind Sports.

One of SCA's objectives was to get to the bottom of what took place in that hospital room at Indiana University Hospital on October 27, 1996.

Almost nine years later, on October 25, 2005, the Texas-based company deposed Frankie and Betsy Andreu. SCA wanted the couple to voluntarily give evidence but they refused and were served with a court order. Betsy Andreu gave her deposition in a Detroit hotel on the morning of October 25, with Lance Armstrong present for the three hours of her examination. Her husband gave his evidence that afternoon. Stephanie McIlvain, another witness to the alleged hospital room admission, was subpoenaed and deposed on November 14, 2005. Jeff Tillotson, one of the lawyers representing SCA, was first to question Betsy Andreu and asked if she had personal knowledge of Armstrong's use of performance-enhancing drugs. Andreu agreed she had. Tillotson then moved on to the critical question of what she had heard in the hospital room.

> **Q:** I want to focus on events that have been alleged in this case regarding certain statements Mr. Armstrong made to doctors. If you can, describe for us now how those events came about, what took place.
>
> **A:** We were in his hospital room, and he was having a scheduled meeting with his doctor. He was having a scheduled meeting with his doctor or with a doctor. I'm not sure. There was going to be some sort of meeting. And there were quite a few people there, so we went into a conference room, and there at the conference room, people that were there were me and Frankie and Lance, Stephanie McIlvain—that was the first time I met her—Chris Carmichael, his then girlfriend now wife, Paige.
>
> **Q:** Anyone else that you recall?
>
> **A:** Lisa Shiels.
>
> **Q:** Who was she?

A: His girlfriend at the time.

Tillotson then asked Andreu if Armstrong's mother Linda was in the room, if his attorney Bill Stapleton was there, if his friend Jim Ochowicz was present? Andreu said none of them were present for the consultation. Andreu was then asked if she knew who the doctors were, but she didn't. Tillotson showed her photographs of two of Armstrong's more high-profile physicians, Dr. Scott Schapiro and Dr. Craig Nichols, and asked if either of them were the doctors. She looked at the photos and shook her head. They were not the ones. Tillotson then got to the point. What happened next?

> **A:** They began to ask him some questions—banal questions. I don't remember. And all of a sudden, boom, "Have you ever done any performance-enhancing drugs?" And he said, yes. And they asked, what were they, and Lance said, EPO, growth hormone, cortisone, steroids, testosterone.
>
> **Q:** Are you absolutely certain that's what he said?
>
> **A:** Yeah, I'm positive.

Betsy Andreu then told how she became upset, turned to Frankie, and insisted they leave the room. Outside, she demanded to know why he hadn't told her Armstrong was doing these drugs and whether he, too, was a drug user. He said he was surprised to hear Armstrong admit using so many performance-enhancing drugs and that he didn't do the stuff that Lance was doing. Her anger came from the thought that she was engaged to a man who might be using performance-enhancing drugs, and she told Frankie that if he was doping, she wouldn't marry him. Andreu said he was not the same as Armstrong. The argument continued all the way back to their hotel room.

> **Q:** You left the room and had this conversation, this difficult conversation with your husband?

A: Yes. And then we went back to the hotel room, and I know I was very upset, and I was very angry.

Q: Why so upset? I mean . . .

A: Because . . .

Q: This is still emotional now. Why was this so upsetting to you? Describe for us why.

A: Because I was going to marry this man, and I am a completely antidrug person, and I was not going to marry somebody who was doing that stuff.

Q: Now, as you recall, you were sitting there when this conversation took place where Mr. Armstrong said the events that you told us today he said. Who else was around you that you believe heard this same thing?

A: Everybody in that room heard it—every single person.

Soon after returning from Indiana to her home in Michigan, Betsy Andreu told a number of people precisely what she had heard in the hospital room. She confided in her mother; her brother; her close friend Dawn Polay, who was going to be matron of honor at the Andreus' wedding on December 31; and a number of other close friends. She recalled speaking with Stephanie McIlvain about the consultation and McIlvain saying how shocked she had been. Andreu said she told the American rider Bobby Julich about the hospital room and also the Parisian-based journalist James Startt. A family friend, Startt was told in an off-the-record conversation.

Andreu also related how she and McIlvain became close friends in more recent years and how they would speak by telephone on a daily basis. She recalled a visit from McIlvain to her home in Dearborn, Michigan, in September 2004, and over the course of her stay, they had discussed the hospital room consultation many times. According to Betsy Andreu, McIlvain fully remembered Armstrong admitting his use of performance-enhancing drugs.

Q: In all those conversations that you had with Stephanie, did she ever once tell you that she didn't believe that Mr. Armstrong had admitted to the use of performance-enhancing drugs in that conversation?

A: No. Stephanie came to visit me and my family, and she, Frankie, and I sat around our kitchen table talking about it. And that's where she said she only remembered Lance saying that he took steroids.

Q: Now, you know in this proceeding she has denied that she heard Mr. Armstrong say such things. Is that true based upon what she had told you previously?

A: She was lying.

Q: I'm sorry, ma'am. I didn't hear you.

A: She was lying.

Frankie Andreu was not comfortable with his involvement in the SCA Promotions–Lance Armstrong litigation. From his perspective, the case could only hurt him and his family. There were no pluses. He worked in a sport that liked to keep its secrets. Doping wasn't a subject for public consumption. Far from being thanked, whistleblowers such as Paul Kimmage, Willy Voet, Jerome Chiotti, Jesus Manzano, and Christophe Bassons were marginalized and eventually forced out of the sport. That's fine if you are ready to walk away from cycling, but Andreu wasn't. The sport had long been his life, and, more important, he wanted it to be his future. But the subpoena had to be answered, even if Andreu did so without enthusiasm. Under oath, Frankie Andreu told much the same story as his wife—except that in his recollection of the drugs mentioned by Armstrong, there were no steroids; just testosterone, r-EPO, human growth hormone, and corticoids.

Asked if he had discussed what he heard in the hospital room with anyone other than his wife, Andreu admitted to relaying the story to his former teammate Bobby Julich, his friend Marty Testasecca, and on one occasion, he spoke with Lance Armstrong about it. Jeff Tillotson was eager to know what was said between himself and Armstrong.

> **A:** He kind of asked about how Betsy reacted to what happened in the hospital room.
>
> **Q:** What did you tell him?
>
> **A:** I said she freaked out a little bit, and, you know, we got into a couple of arguments, but then it kind of went away.

Five weeks after the Andreus were deposed in Michigan, Lance Armstrong made his way to a ten o'clock appointment at the offices of Herman Howry & Breen, 1900 Pearl Street, Austin, Texas. It was the last day of November 2005, and the morning of his deposition in the case involving SCA. After explaining the rules of play at the tribunal, SCA's lawyer Jeff Tillotson soon focused on what took place in the hospital conference room.

> **Q:** Okay. Do you have any recollection while these individuals [the visitors] were there that a doctor or doctors came into the room and discussed with you your medical treatment or your condition?
>
> **A:** Absolutely not.
>
> **Q:** Okay.
>
> **A:** That didn't happen.
>
> **Q:** Did any medical person ask you, while you were at the Indiana University Hospital, whether you had ever used any sort of performance-enhancing drugs or substances?
>
> **A:** No. Absolutely not.

Q: So that just never came up. No one ever, as part of your treatment, no one ever asked you that?

A: No.

Q: Can you offer, or can you, can you help explain to me why Ms. Andreu would make that story up?

A: Well, she said in her deposition she hates me.

Q: Do you believe she's making that story up to get back at you or to cause you harm?

A: Whether she's making up that she hates me?

There are two points worth considering. First, Armstrong's insistence that doctors did not come into his room and discuss his medical treatment, and second, his recollection that Betsy Andreu admitted in her deposition that she hated him. Armstrong remembered having visitors that Sunday afternoon and recalled the move to the bigger conference room with the Cowboys' football game on television. But, according to him, no doctors entered. "That didn't happen." On the other hand, Timothy Herman, his attorney, seemed to accept two doctors were in the room and that their conversation with the patient led to a misunderstanding that might explain everything. In his cross-examination of Betsy Andreu, Herman asked: "Do you remember any conversation about treatment, postoperative treatment involving steroids?"

A: With Lance?

Q: Right.

A: No.

Q: You don't recall the doctors talking about that?

A: No. Because I wasn't there for what his treatment was going to be afterward. We just wanted to make sure he was going to live.

Q: Well, were you aware that he had had a regimen of r-EPO that was part of his treatment for cancer? Were you aware of that?

A: As most cancer patients, yes.

In a later interview with National Public Radio, Herman told the broadcaster Tom Goldman, "Mr. Armstrong was taking steroids at the time, as part of his postoperative treatment. . . . It's very possible there could've been mention of steroids and r-EPO in this conversation with these two doctors indicating either the current regimen, or the regimen that Armstrong was gonna be subject to after this surgery, or when he got out of the hospital."

As for Armstrong's contention that Betsy Andreu admitted in her deposition that she hated him, that simply did not happen. The exchange Armstrong referred to involved Herman and Betsy Andreu. During the deposition, the lawyer recalled the conversation between her husband and Bill Stapleton and Stapleton's associate Bart Knaggs at the 2004 Tour de France. "And without going through in detail," said Herman, "Frankie, on several occasions in that conversation with Mr. Knaggs and Mr. Stapleton, talks about how much you hate Lance Armstrong, doesn't he?"

A: No. He just says they don't like each other, but that's the extent of it.

Q: Okay. So, is it true or untrue that you do not like Mr. Armstrong?

A: No. I don't like him.

If something is bothering Lance Armstrong and he can deal with it through a telephone call, it seems he doesn't think twice before dialing the number. Greg LeMond got such a phone call after his comments questioning the rider's association with Michele Ferrari in

2001. The former U.S. Postal Service team doctor Prentice Steffen had another such call after Armstrong presumed he was the doctor raising suspicions about the champion in the July 2001 edition of the magazine *Texas Monthly*. Both LeMond and Steffen believe the calls were threatening and intended to silence them. Armstrong denies this. So fraught was the conversation between Armstrong and her husband that Kathy LeMond made notes while it was happening; Prentice Steffen also wrote a short account of his vexed conversation with Armstrong immediately after it happened.

Armstrong also called up witnesses deposed by SCA and spoke to them by telephone shortly before they gave evidence. Frankie Andreu, the Paris-based journalist James Startt, and McIlvain were called in this manner. Frankie Andreu received his phone call the day before he gave evidence. Armstrong was asked about the call to Andreu by Jeff Tillotson.

Q: Now, prior to Mr. Andreu's deposition, you did call him, did you not?

A: I . . . yes.

Q: Did you actually speak to him?

A: Yes.

Q: What was your reason for calling him?

A: Well, I think I called because . . . because we . . . because Kathy LeMond had done her deposition, and had all kinds of crazy things to say, which were news to us.

Q: Any other reason you called him?

A: Other than to say hello, no.

Q: Were you trying to influence his testimony in any way?

A: Of course not.

In his deposition, Andreu was asked about the same phone call. He explained that much of their conversation concerned Kathy LeMond's deposition and Armstrong's insistence that her allegations against him weren't true. Andreu in his deposition was then asked if Armstrong had brought up the hospital room during their phone call, and Andreu replied that Armstrong had. According to Andreu, Armstrong told him the hospital room confession never happened. Believing there was no point in arguing, Andreu chose to remain silent. This was a subject they had discussed a long time ago, when Armstrong asked him how Betsy reacted to what she had heard in the hospital room.

Tillotson also asked Frankie Andreu if he considered it odd that Armstrong should have denied the hospital room in that telephone conversation.

> **A:** I considered it odd that he even called me, because I hadn't spoken with Lance in probably two and a half years.

According to the Andreus and Stephanie McIlvain, six visitors were present in the conference room at Indiana University Hospital when the alleged admission of drug-taking was made. SCA served subpoenas on Frankie and Betsy Andreu and McIlvain. That left Chris and Paige Carmichael, and Armstrong's onetime girlfriend Lisa Shiels. SCA's representatives tried unsuccessfully to contact Lisa Shiels. They thought about the Carmichaels but expected them to appear as witnesses for the other side. Chris Carmichael was Armstrong's principal trainer until the arrival of Michele Ferrari, and he had remained his friend and business associate. Though the Andreus and McIlvain agreed the Carmichaels and Lisa Shiels were in the room during the consultation, they did not appear as witnesses in the case.

Much depended upon the testimony of Stephanie McIlvain.

She worked for the sportswear company Oakley as the company's liaison with Armstrong and over the years she and the cyclist became good friends. Before deciding to make Bill Stapleton his manager in late 1993, Armstrong asked McIlvain for her opinion on the prospective candidates. In 2001, after her two-year-old son Dylan was diagnosed with autism, McIlvain decided to stop working but Armstrong intervened and made it possible for her to work from home while looking after her son. Contacted by one of the authors of *L.A. Confidentiel* in early 2004, McIlvain admitted being in the hospital room when Armstrong had the contentious consultation with his doctors, and in an on-the-record comment, she refused to say she did not hear Armstrong confess to using performance-enhancing drugs. Instead, she apologized and said it was a question for the rider to answer.

Having met for the first time in that hospital room, Betsy Andreu and Stephanie McIlvain became such good friends that by 2004 they were speaking daily by telephone. According to Andreu, they discussed the hospital room incident many times and were surprised the story had remained untold for so long. On the question of what precisely took place in the hospital room, Stephanie McIlvain's evidence was very important. She worked with Armstrong but was friends with the Andreus. They said Armstrong confessed to doping, he said it wasn't true; Betsy Andreu, Armstrong said, made it up and her husband felt he had to support her. The Andreus were in no doubt that McIlvain heard what transpired. At McIlvain's deposition, Timothy Herman was first to question her.

> **Q:** In connection with your visits to the hospital with Mr. Armstrong, did there ever come a time where you were with him and with other people where there was any discussion regarding Mr. Armstrong's use of performance-enhancing drugs or substances?
>
> **A:** No.
>
> **Q:** Okay. There's been testimony. . . . Let me rephrase that. Were you ever at a hospital room or other part of the hospital with Mr. Armstrong where he said anything about performance-enhancing drugs?

A: No.

Q: Do you have any recollection of any doctor in your presence asking Mr. Armstrong if he used in the past any performance-enhancing drugs or substances?

A: No.

Q: Okay. Did you talk with anyone at the Indiana University Hospital or its surroundings about Mr. Armstrong's use of performance-enhancing drugs?

A: No.

Q: Since Mr. Armstrong's treatment, have you ever spoken with any other person about whether or not Mr. Armstrong told someone at the hospital that he used performance-enhancing drugs?

A: Yes.

Q: Who have you spoken to?

A: I spoke with Betsy Andreu and Frankie.

Q: Do you remember when that was?

A: Just probably about four years ago.

Q: Tell me first what was the occasion why you were talking to them and then I'm going to ask you what you talked about.

A: Betsy Andreu called up and asked if I remembered an incident where two doctors came in and Lance told them what drugs he had used, and I, at that point, I told her no, I don't. I don't remember Lance ever saying to two doctors that he used drugs. I do remember being in a conference room or a room with some people, and Betsy and Frankie were in there, and I came in, and the reason I remember it so well is because they were watching a football game and I—sorry,

> everybody—but I hate football and it was—sorry—it was very painful for me, so I went and sat down on the floor where the couch was and I just sat there and watched, watched the football game, and that's the main thing that I remember, but when Betsy called me and talked to me about it, she started telling me what she heard and what she saw in that room.
>
> **Q:** What is it she told you?
>
> **A:** She said that she remembers being in that room and she swore that I was in that room and that I had to hear it, that two, two gentlemen came in and asked him what he was using, and at that point, he told them whatever it was.

But McIlvain insisted she didn't hear it and, consequently, could never have told anyone she did hear it. This contradicted the evidence of her friend Betsy Andreu and raised the obvious question of who was telling the truth. McIlvain was also asked if she had spoken to Greg LeMond about the hospital room, a line of questioning that followed on from LeMond saying in his deposition that McIlvain had told him about it.

> **Q:** Is there no truth to the statements that you told Mr. Greg LeMond that you were present at the Indiana hospital room and confirmed what Betsy Andreu told us, which is that Mr. Armstrong admitted to EPO use, growth hormone, testosterone, and other drugs?
>
> **A:** No. I told Greg LeMond I remember being in a room and I remember watching a football game and first meeting Betsy and Paige Carmichael.
>
> **Q:** Do you remember if Mr. LeMond asked you if Mr. Armstrong said he used drugs while you were in that room?
>
> **A:** He . . . He told me what Betsy told him and asked me if I remember it that way.

Q: And your response to Mr. LeMond was?

A: No. I remember being in a room.

News of McIlvain's testimony swept through the small circle of people involved in the SCA–Tailwind Sports litigation. Greg LeMond was one of those who heard that McIlvain had denied hearing anything in the hospital room. This upset him because McIlvain had told him she was present and did hear it. The Andreus were not pleased because they were now on their own as the only witnesses to a damning and disputed allegation. They believed their friend had perjured herself. Betsy Andreu felt let down while her husband was more sympathetic, pointing out that McIlvain worked for one of Armstrong's sponsors, Oakley, and her job depended upon having a relationship with the rider. McIlvain's evidence did not help SCA's case.

SCA Promotions was aware from Kathy LeMond's deposition that her husband, in an effort to protect himself, taped several telephone conversations relating to his contretemps with Armstrong. Greg LeMond's reasoning was straightforward; he was convinced his bicycle business was in danger because of his public comments about Armstrong, and thinking he might need proof of the pressure from powerful businessmen, he taped some telephone conversations. SCA knew of a taped conversation with McIlvain and issued a subpoena that obliged LeMond to put it before the tribunal.

It is legal for citizens in Minnesota, where LeMond lives, to record telephone conversations without the permission of the other party. However, it is not legal in every U.S. state and it is illegal for a Minnesota resident to tape a call with a person from out of state if the laws in the other state are different. LeMond claims not to have known this when he called McIlvain on July 21, 2004, at her home in California and taped the conversation. California law forbids this practice. The resulting tape, though, is an extraordinary conversation that goes on for more than thirty-two minutes and offers an insight into the thinking and feelings of a woman who had worked with Armstrong for more than twelve years, and had been a close friend of his, and was still working with him at the time of the call. The three-man panel of arbitrators and the various lawyers present were aware McIlvain had denied hearing Armstrong say anything about performance-enhancing drugs in the hospital room. When the tape was played, they heard her tell a dramatically different story. The question of what McIlvain heard in the hospital room arose when LeMond suggested he might need her to testify in a possible future lawsuit.

LeMond: So, and I, I know, I mean I know what I heard from a source outside of the group here of what happened at the hospital, and Betsy and I have talked a little bit, but, and I, I'm not asking you to do anything you would never want to do, but, you know, if I did get down where it was a . . . you know, a lawsuit, are you, would you be willing to testify?

McIlvain: If I was subpoenaed, I would.

LeMond: Yeah.

McIlvain: 'Cause I'm not going to lie. You know, I was in that room. I heard it.

A little further on in the conversation, LeMond speculates on the consequences of Armstrong falling from grace and the problems he would face dealing with that. But McIlvain isn't sure there will be a fall.

McIlvain: Well, the whole thing of it is, Greg, there is so many people protecting him that it is just sickening, you know.

LeMond: But the people protecting him know.

McIlvain: I know they all know.

LeMond: Yeah.

McIlvain: Well, because I know, and this you don't repeat, but I know, for a fact, that when that whole book came out, Chris

> Carmichael made a call to my friend and said, "Oh, you know, I've been sitting here, thinking, thinking, thinking who was in that room. If I totally remember the incident, yes he did admit to what he was taking. But I don't really think Stephanie and Betsy Andreu were in there, and I don't think Lisa Shiels was in there." And I just laughed. I said, "You tell him that yeah, I was in there, because I remember him [Chris] looking around the room and seeing who was in that room." So then my friend says, "Oh my God, that's what he said. He said he looked around the room to make sure everybody that was in that room could be trustworthy."

LeMond again brings up the possibility of fighting a lawsuit and needing people to support him. As in the earlier part of the conversation, McIlvain reassures him she won't lie under oath. "Being subpoenaed," she says, "I don't have a problem because I already told Jim . . . I told him I won't lie. [The 'Jim' that McIlvain refers to is Jim Jannard, founder of Oakley.] 'And,' he says, 'well, you know, there's certain ways to get around it.' I said, 'I don't care, Jim.' I said, 'I won't lie.' I said, 'I wouldn't be able to live with myself.' You know, because I, too, know quite a bit, 'cause Lance and I were close, you know. But, um, I definitely wouldn't lie about that [the hospital room] because it's public knowledge. A lot of people know about it, you know. And the part that pisses me off about the whole thing, even if we were close right now, it's how many people he has given false hope to, and I think that is the most disgusting thing ever for someone to do. . . . From somebody who has a child with a handicap, you look up to people who've gone through the same thing, and you look for hope and you look for strength and for him to be doing that to those poor people who look up to him and honestly think that he's doing this 'cause he's a superman and . . ."

> **LeMond:** I agree. I had . . .
>
> **McIlvain:** It kills me.

On a separate occasion, the journalist James Startt testified under oath that he discussed the hospital room with McIlvain and she had told him about hearing Armstrong admit to his doctors that he had taken banned drugs.

In his sworn deposition on November 30, 1995, Lance Armstrong emphatically denied that he had admitted taking performance-enhancing drugs during a consultation at Indiana University Hospital. Confronted with the accounts of both Betsy and Frankie Andreu, he refuted them: "100 percent, absolutely." He pointed out that he could not remember being in the room with the six visitors listed by the Andreus and said he found it interesting that some people were missing: his close friend Jim Ochowicz, his attorney Bill Stapleton, his mother Linda, his friend John Korioth. He couldn't imagine that those people weren't present during every visitation and he said it would have been impossible for him to be anywhere in the hospital without Ochowicz and Stapleton. It was untrue, he said, that a doctor or doctors discussed his medical treatment while visitors were in the room and also untrue to say doctors at Indiana University Hospital asked if he had ever used performance-enhancing drugs. The only part of the Andreus' account that he recalled was the Dallas Cowboys' game on TV.

Armstrong believes malice lay at the heart of Betsy Andreu's evidence. Because she disliked him, she concocted a viciously damaging story. From there, he believes it is easy to explain Frankie Andreu's corroboration of his wife's account, a perjury so understandable that Armstrong said he felt for the guy. "Frankie is trying to back up his old lady." There is an obvious difficulty with this theory, and it is timing. The hospital room conversation happened in 1996, three years before the first hint of disharmony in the Armstrong–Betsy Andreu relationship. From '97 through '99, the Andreus were close friends with Armstrong and his then girlfriend, Kristin Richard. Betsy got to know Lance when she visited Frankie in Como, and later both couples spent time together when they were all living in Nice. Betsy, a devout Catholic, and Lance, a staunch atheist, would argue over religion and if Lance was at a loose end, he might call up Betsy and ask her to come round and cook his favorite risotto.

If malice was the reason behind Andreu's allegation, it follows that the hospital room story must have been fabricated at a time when some bad feeling existed between her and Armstrong: after 1999. But the possibility of the story being concocted three or four years after the event withers in the light of the evidence. Betsy Andreu was so upset by what she heard in Indiana that immediately after her return to Dearborn, Michigan, she told many people about what she'd heard. Vivian Hackman was one of the friends in whom Andreu confided. They grew up in the same neighborhood in Michigan and were lifelong friends. "Betsy was still in Indiana when she called me, and she said, 'I will talk to you when I get home but I might not be getting married.' She came home and she was really upset. She told me they were in the hospital room, the doctors came in, she wanted to leave, and Lance said it was okay to stay. Then he is listing all these drugs and Betsy just couldn't believe it. She had looked up to Lance but he admitted using performance-enhancing drugs and in her eyes that was wrong. She didn't agree with it and if Frankie was doing the same, she wasn't going to marry him."

Another friend, Lory Testasecca, recalls a similar conversation. "We were in Ann Arbor somewhere, having a coffee, just me and Betsy. It had to have been late October or early November 1996, because they got married in December. I could tell she wanted to tell me something but she didn't want to tell me and finally she did. She said she and Frankie were in the hospital room in Indiana and they heard the doctor ask him [Armstrong] if he had ever taken any performance-enhancing drugs and he said yes. I felt she kind of wished she didn't hear it but she did." Piero Boccarossa has been a friend of the Kramar family for twenty years and is particularly close to Betsy. Soon after her return from Indiana, Betsy told him what she had heard. He was concerned and said if Frankie was doing the same, she shouldn't marry him. A relative of his, who was a bodybuilder, had used steroids and he had seen the effects.

Another close friend of Andreu's, Dawn Polay, remembers how upset Betsy was by what had happened in Indiana. Polay was pregnant with her first child at the time; Betsy would be godmother and Polay would be matron of honor at Betsy's wedding. She remembers they spoke continuously about what happened in the hospital room. "Betsy hadn't been aware of what Armstrong was doing. She didn't want to be involved with anybody who did this and she threatened to call off her wedding to Frankie. The thing about Betsy is that she's very by the book. She plays by the rules, black and white, there's very little gray with her."

Armstrong's oncologist at Indiana University Hospital, Dr. Craig Nichols, signed a sworn affidavit that was submitted during the SCA–Tailwind Sports arbitration disputing the claim that his patient admitted using performance-enhancing drugs. "He never admitted, suggested, or indicated that he has ever taken performance-enhancing drugs," said Nichols. "Had this been disclosed to me, I would have recorded it, or been aware of it, as a pertinent aspect of Lance Armstrong's past medical history. Had I been present at any such 'confession,' I would most certainly have vividly recalled the fact. I would have recorded such a confession as a matter of form, as indeed, would have my colleagues. None was recorded.

"Though doctors are under a professional obligation to record all matters regarding a patient's medical history in his/her notes, it would be unusual to ask a professional athlete who has been diagnosed with testicular cancer whether or not he has used performance-enhancing drugs. I have treated other athletes with testicular cancer and don't recall ever asking them whether or not they have used performance-enhancing drugs."

Betsy Andreu said Nichols was not one of the two doctors who entered the room that Sunday evening. It is also interesting that Nichols said it would have been unusual for any doctor to ask a professional athlete with testicular cancer whether he had used performance-enhancing drugs. Other oncologists disagree with him on this point and Armstrong's lawyer, Herman, told the Associated Press that doctors in Indiana repeatedly asked Armstrong about substances he may have used and were told only that he occasionally drank beer. Nichols, who is a member of the Scientific Advisory Committee of the Lance Armstrong Foundation, is now the chair of hematology-oncology at Oregon Health and Science University.

Two days after Frankie and Betsy Andreu first testified about the hospital room incident during their depositions on October 25, 2005, the Lance Armstrong Foundation announced it was establishing an endowment in oncology at Indiana University Hospital, at a cost of $1.5 million. During the SCA process, Jeff Tillotson pointed out the coincidental timing of the payment. "Can I say one thing real quick," replied Armstrong. "A donation like this takes a lot of time and a lot of thought and there was a major process. This is not a spontaneous onetime, hey, let's cut the check to somebody that we hardly know. This may be the date that this was issued . . . but Dr. Einhorn, he's too legendary to go there on this. This is a man of the highest standard. You're not the president of ASCO [American Society of Clinical Oncology] for taking payoffs, ever. That's as prestigious as it gets in the world of American oncology, so I just want to be clear about that."

Chapter 19

ONE IN A BILLION?

Dr. Mike Ashenden is a slim, neatly built Australian. One look and you'd guess he's a long-distance runner. He has done triathlons but only for pleasure. Physiology and sports science are his business. His primary university degree was in exercise and sports science. His Ph.D., from James Cook University in north Queensland, was a study into the effects of altitude on the blood of elite athletes, especially on the production of red cells. While preparing his doctoral thesis in the late 1990s, Ashenden worked in the Department of Physiology at the Australian Institute of Sport (AIS) in Canberra and came into contact with many athletes—sufficient to convince him doping was the poison with the potential to destroy professional sports. Perhaps unknowingly at first, Ashenden became one of those people who swapped a job for a vocation. Anti-doping became his life. He was part of the Sydney 2000 team that produced a test to detect r-EPO in blood. In 2001 he set up an international research group called Science and Industry Against Blood Doping (SIAB) and coordinated a project that delivered a test to detect blood substitutes. Under the SIAB banner, he coordinated the research into the applicability of the homologous blood test used to catch the cyclists Tyler Hamilton and Santiago Perez in 2004. The cyclists were banned for two years each.

In September 2004, after the two riders' positives, Ashenden expressed his feelings about blood doping on the website cyclingnews .com. "It is," he said, "the most diabolical form of doping. . . . It has to be stamped out, and I hope that if there is one positive that comes out of it [the Hamilton and Perez positive tests], it is that athletes using this form of doping know they have to stop. It's repugnant."

Seven months later, Ashenden received an e-mail from Chris Compton, an in-house attorney with SCA Promotions. Compton wanted to know if Ashenden would be interested in helping SCA with its case against Lance Armstrong. Specifically, Compton wanted the Australian anti-doping expert to interpret blood profiles belonging to the cyclist. Part of what Ashenden did for a living was assess longitudinal blood profiles to distinguish dopers from clean athletes. He thought he might be able to help Compton and they had further conversations. Following increased telephone and e-mail contact, the lawyer asked Ashenden if he would be willing to act as SCA's expert witness in the case.

So it happened that Dr. Mike Ashenden became involved in *Lance Armstrong* v. *SCA Promotions*.

Ashenden's experience at the Australian Institute of Sport familiarized him with the damage caused to sport by doping. Aussie athletes would return from international competition complaining they couldn't compete because international rivals were doping; they would name names, and even if Ashenden's natural skepticism made him wary of the allegations, he couldn't deny that in many sports, drug cheats were prospering. As a scientist, he knew more could be done to combat the problem.

Offered the chance to work on the Armstrong case, Ashenden's instinct was to accept. The cyclist interested him. Not because the doctor knew more about the Texan than the average fan (he didn't) but because Armstrong epitomized professional sports' great dilemma at the start of the twenty-first century: you didn't know he cheated but you couldn't be sure he didn't. Before that first e-mail from Chris Compton, Ashenden had a fan's interest in the Tour de France. He was aware of Armstrong's dominance and, like everyone else, knew about the recovery from life-threatening cancer. He had heard, too, the suspicion that the rider doped, and he remembered a French police investigation into Armstrong and his team. As well as that, he recalled Armstrong worked with a shady Italian doctor.

What intrigued him most about Armstrong was the transformation from an okay Tour de France rider to an extraordinary one. When he read stuff about the rider, he paid particular attention to the parts explaining why, after his treatment for cancer, he was able to achieve victories that had not seemed possible before his illness. No matter how much he read, Ashenden wasn't convinced. Armstrong lost weight, he became more dedicated, he trained smarter, he ate healthier, he had become more mature, he pedaled more efficiently, he focused more on the Tour—all of these things would definitely help. But could they bring about that huge transformation? The scientist was unsure. It wasn't that Ashenden believed Armstrong was doping but that he couldn't be sure he wasn't. That same old story. So, yeah, he told Compton, this was a case he would like to explore.

Timothy Herman and Sean Breen sat in an office in Austin, Texas; Jeff Tillotson and Chris Compton were waiting in an office in Dallas; Mark Levinstein was in another law office in Washington, D.C., and in faraway Australia, Mike Ashenden was ready for the lawyers' questions. It was December 2005; three days before Christmas Day in the United States, two days before Christmas for Ashenden down under, and they were linked by conference call to take the Australian's deposition. Armstrong's legal team chose Levinstein to ask the questions of Ashenden because he was experienced in sports law and had an understanding of the science and protocols of drug testing. More than eight months had passed since Ashenden had received that first e-mail from Compton, and over the intervening time, he had tried to piece together the extraordinary story of Lance Armstrong. The more time Ashenden spent on the case, the deeper his suspicions became.

Depositions are taken to elicit from a witness what testimony he or she intends to offer during the actual case. Levinstein quickly got to the key questions: was the expert witness going to say Armstrong had used blood transfusions? Ashenden replied that the '99 results were clear evidence of wrongdoing and that, in later years, Armstrong could have transfused blood and not been detected. He then referred to three hematocrit readings for Armstrong quoted in the book *L.A. Confidentiel* and said they supported the view that he could have used blood transfusion. Levinstein wondered if there was any other evidence suggesting Armstrong cheated.

"His coach, Chris Carmichael," said Ashenden, "was a member of the U.S. cycling team that used both homologous and autologous blood transfusions [for the 1984 Los Angeles Olympics], so his coach has firsthand knowledge of both of those approaches."

When Levinstein asked if there was anything else, Ashenden quoted Armstrong's collaboration with Michele Ferrari. He also referred to U.S. Postal Service urine samples from the 2000 Tour de France that French investigators described as being "too clear"—the scientists examining the urine had been puzzled that after a long stage of the Tour de France, Postal riders should have produced crystal clear urine. From the tone of his questions, it was clear Levinstein was unimpressed by Ashenden. He felt the associations with Ferrari and Carmichael were public knowledge and that one didn't need to be an expert witness to comment on that. What else was there—the story of six alleged positives for r-EPO as reported in *L'Equipe*, three hematocrit readings for Armstrong in *L.A. Confidentiel* that anyone could see, and urine samples from the 2000 Tour de France that were "too clear." It didn't seem to amount to much.

Almost four weeks later, on the afternoon of January 18, 2006,

Dr. Mike Ashenden took the witness seat in the arbitration of the *Armstrong* v. *SCA* case at the offices of Richard Faulkner on North Central Expressway in Dallas. Following ten days of listening to the evidence presented by various witnesses and months of sifting through background material related to the case, Ashenden was ready to offer his judgment on one of the most vexing sport stories of our time.

Cody Towns, a lawyer representing SCA, began the questioning of Ashenden. Forty-four minutes later, Towns asked the question that towered over the arbitration from the moment it began. "Do you have an opinion on whether Lance Armstrong used performance-enhancing drugs?"

> **MA:** Yes, I do.
>
> **CT:** And what is that opinion?
>
> **MA:** Based on what I've seen and read and heard, to my mind, beyond any reasonable doubt, he has used performance-enhancing drugs at some point.
>
> **CT:** And what specifically leads you to that conclusion?
>
> **MA:** I think it's a conglomeration of things. The thing that I've always been aware of, that never really made sense to me, was the sudden jump in performance. Now, in anti-doping research, one of the things you look for is a sudden unexplained improvement in the performance.

Towns then asked Ashenden what concerned him about the dramatic improvement in Armstrong's performance.

> **MA:** The thing that concerns me is that pre-cancer, the Tour results in '93 and '94, for example, [where] he struggled to keep up with the peloton, especially in the mountain stages. He was dropping twenty minutes at a time.

Ashenden's argument that Armstrong didn't compare with the best in any of his first four Tours is borne out by the results. The American made his debut in the 1993 Tour and rode just two mountain stages before abandoning. On the 203-kilometer race from Villard-de-Lans to Serre Chevalier in the Alps, he finished 86th, 21 minutes and 42 seconds behind the stage winner, Tony Rominger. The following day was tougher, a 263-kilometer ride to the ski station at Isola 2000, where Armstrong came in 97th, 28 minutes and 47 seconds down on Rominger, who was winning his second consecutive stage. On both days Armstrong rode hard, without overextending himself to finish a few places higher. No one gave it a second thought when he quit the race the next morning; he was a twenty-one-year-old who had come to that first Tour to learn and then depart before the lessons grew too painful. Some improvement in the mountains might have been expected a year later but it wasn't forthcoming. In 1994, Armstrong finished 64th on the Cahors-to-Hautacam stage in the Pyrenees, 7 minutes behind the stage winner, Luc Leblanc, and then 55th on the following day's race to Luz Ardiden, 20 minutes down on that day's winner, Richard Virenque.

His third participation in 1995 was the most revealing. By now Armstrong was twenty-three, and with the experiences of '93 and '94, he was ready for the twenty-three-day marathon and was in the best form of his career. In his evidence to the court of arbitration he said, "You know, for me personally, 1995 was a good year. I was riding very well, I was successful in spring, I won the Tour DuPont, won a stage of the Tour de France . . . I felt great." It would be the only Tour de France that Armstrong completed pre-cancer. He finished 36th, 1 hour, 28 minutes, and 6 seconds down on the Tour winner, Miguel Indurain. It was a good effort from Armstrong but when it came to the mountains, he just didn't have it. He completed all five mountain stages, achieving his best place in the least difficult leg from Saint-Etienne to Mende, 39th—and 8 minutes and 37 seconds behind that day's winner, Laurent Jalabert. Otherwise, it was sustained and unavailing hardship; 40th and almost 18 minutes behind Alex Zulle on the Alpine stage to La Plagne; 56th and 18.44 minutes down on Marco Pantani on the following day's leg to Alpe d'Huez; over 28 minutes behind Pantani in 117th place on the Pyrenean stage to Guzet Neige; 64th and almost 33 minutes behind Richard Virenque at the summit finish in Cauterets.

"In today's tour," the eminent French writer Pierre Chany once said, "you see tomorrow's winner." In Armstrong's 1995 performance, there wasn't a hint as to what was to come.

New York born and raised, Ed Coyle is a professor at the University of Texas at Austin, where he is director of the human performance laboratory. He also teaches, publishes articles in various journals, trains doctoral students, and conducts physiological tests on athletes. In November 1992, the then twenty-one-year-old Lance Armstrong walked into Coyle's laboratory in Austin and did a battery of tests. On four further occasions, Armstrong was tested by Coyle, culminating in a series of tests in November 1999—four months after Armstrong won his first Tour de France. In March 2005, Coyle published a study on Armstrong in the *Journal of Applied Physiology* that attempted to show how a four-time noncontender became the greatest racer in the history of the Tour de France. Because Coyle's data on Armstrong is the only data that has been publicized, it is important. Coyle called his paper "Improved Muscular Efficiency Displayed as Tour de France Champion Matures" and he concluded it thus: "Clearly, this champion embodies a phenomenon of both genetic natural selection and the extreme to which the human can adapt to endurance training performed for a decade or more in a person who is truly inspired."

On January 13, 2006, Coyle appeared as a witness for Armstrong in the case against SCA Promotions, Inc. He was first questioned by Lisa Blue, an attorney on behalf of Armstrong, and he quickly set out his central argument: "Simply," he said, "this boils down to somebody who very young had a lot of raw, natural talent, and I'm talking about physiological talent here, which is not psychological, but a huge cardiovascular system or ability to produce energy aerobically, just raw power. And I think he kind of reflects [this] in some aspects of life. That is, as he got older and matured, he became more efficient and just learned how to more wisely apply that power to the purpose which he directed to winning the Tour de France. So just a lot of raw energy, and then he became more efficient, his muscles actually became more efficient at transmitting that raw energy to powering the bicycle and simply he improved his power-to-body-weight ratio a remarkable 18 percent."

Clearly impressed by this summary, Lisa Blue asked Ed Coyle to stop right there and repeat what he had just said:

"Well, over the seven-year period that we tested him, his formative years from age twenty-one through twenty-nine, he showed a remarkable improvement in how much power he can generate for every kilogram of body weight. He went up 18 percent and that's because he both improved his raw power ability by 8 percent because of muscle adaptations which we ascribe to just the pure result of hard training." Coyle then explained the second part of the Armstrong equation: "And then he lost body weight and the two combined equally to cause this 18 percent increase in power, and therefore speed with which he could ride up the mountains in France."

Coyle listed some of the factors behind Armstrong's prowess. "I mean, just genetically he was probably born with a large heart. . . . He's probably, you know, a five-foot-ten individual who naturally has a heart the size of a person who's six-foot-six, and he grew it to a heart the size of a person who's seven-foot-six." This observation was part of a PowerPoint presentation that Coyle made to the arbitration panel. During the presentation, he advanced his theory that Armstrong increased his power output by changing his muscle fiber composition—that is, he converted fast-twitch muscle fibers into slow-twitch muscle fibers, more suitable to the endurance athlete. Coyle did not offer any data to support this.

Toward the end of his presentation, Ed Coyle tried to explain that Armstrong didn't have to be a genetic freak to do what he did, because Armstrong was one in a billion. He put up a slide and explained:

"Well, this slide, again, is a mathematical expression. And that is if Lance were, you know, Lance is the best endurance athlete on the planet. You say, how is that possible? You could say, as the press has said, that he's one in a billion on the planet. What does that take? Does he have to be a genetic freak or superhuman in any one compartment? My point is you don't have to. You just have to make sure you don't have a weakness, okay, that you identify your weakest link and you improve that as much as possible, and that's what I believe Lance has accomplished.

"My point here is if we take just these factors and add them up as to what it takes to be one in a billion, you can see all these lines are connected. My point is, it's a probability statement. A person who is a competitive bicyclist who trains for a number of years only has to be one out of ten in muscle capillary density—blood vessels around the muscle that remove the lactic acid; one out of twenty in stroke volume—heart size and how much blood they can pump; one out of two in hemoglobin content; one out five in the mitochondria—or the aerobic enzymes where the raw energy is produced; the technique of bicycling, one out of ten; and then percentage of slow-twitch one out of five hundred, which I believe is very important for being efficient, and Lance's efficiency certainly is raised from being very average or below average to being superior from years of hard training. We have documented that.

"The point is, if you multiply ten times twenty times two times five times ten times five hundred, the end product of that is one in a billion. That's all it takes. You don't have to be a genetic freak." Neither did one have to be a math freak to work out that Professor Coyle's numbers multiplied came to ten million, not a billion.

Mike Ashenden listened to this "statement of probability" and considered it nonsense.

Though they appeared as witnesses at different times in the *Armstrong* v. *SCA* litigation and never conversed directly, Ashenden and Coyle disagreed on pretty much everything. The irony is that Ashenden's criticisms were based largely on the data collected by Coyle from his five testing sessions with Armstrong. Coyle believed Armstrong to be physiologically gifted, even by the standards of his rivals in elite cycling. But one indication of an athlete's physiological prowess in an endurance sport is the VO_2 Max, a figure that represents the body's capacity to burn oxygen. It is a simple equation: the more oxygen an athlete can supply to the muscle, the more oxygen the muscle can consume. In his paper for the *Journal of Applied Physiology*, Coyle estimated Armstrong's VO_2 Max was higher than that of five-time winner of the Tour de France Miguel Indurain and reckoned Armstrong's number to be among the highest values reported in world-class runners and cyclists.

In his evidence, Ashenden was asked what he saw in the data produced by Coyle on this question: "I guess when I first saw this article [published in the *Journal of Applied Physiology*], the thing that struck me is if you have a look at the maximal oxygen uptake values [VO_2 Max]—you know, seventy, seventy-six, eighty-one, sixty-six, seventy-one—those aren't the sort of values that I would have expected to see in an athlete who could literally leave behind the best cyclists on the planet. So it struck me as, gee, that's lower than what I would have expected." Speaking five months after the case ended, Ashenden elaborated on his point about Armstrong's VO_2 Max. "From the figures in the Coyle study, a fair estimate of Armstrong's VO_2 Max would be somewhere in the midseventies. To test someone five times and pick the highest number as the VO_2 would be akin to a golfer playing five rounds of golf, taking his best score, and claiming that was his level."

SCA's lawyer Cody Towns asked Ashenden if the VO_2 Max figures, as listed by Coyle, could explain Armstrong's improvement in performance. "Well, no," said Ashenden. "The highest value you see there is 81.2 in the middle column, and he—that was at the time that Lance Armstrong won the world championship [September 1993]. Now, 81.2 is a good value. There's no doubt about that. It's a good value. And if an athlete came into your laboratory and had a VO_2 of 81.2, you'd have no hesitation whatsoever in predicting they would have a successful career in an endurance sport. There's no doubt about that. The thing that concerns me there is that that's the highest value that we see, and it goes down from there. It goes to 71.5 in '99. Now, that's inconsistent with the performances. The [Tour de France] success began in '99 when the VO_2 was just 71.5.

"Now, when you start looking in the low seventies like that, your average professional cycling team is going to have at least several athletes with VO_2s equal to 71 and probably higher. I mean, there's a paper that's been published on a Spanish professional cycling team, and the average VO_2 of the entire squad was something like 78.8. That's the average, so obviously some of the cyclists were higher, some of them were lower."

A copy of a page from that study was blown up on a screen for the arbitration panel and Ashenden went through it to emphasize his point. "The highest value that they had in their squad was 84.8. Now, of course, 84.8 is a very good score, but to my mind, to take a message from that is that a VO_2 of—the low seventies, it's still good—but these are the caliber of cyclists that Armstrong would have been competing against, and if his VO_2 when he raced was in the low seventies, he was racing against people who had higher VO_2s than he did. So, in and of itself, the maximum oxygen uptake doesn't explain his success. And then you bring that back and you look at the inconsistency where it was higher in '93 when he didn't have the same success, lower in '99. You really begin to question that. I don't buy the argument that it was some—I think he [Coyle] called it in the article 'exceptionally high VO_2'—that could explain the success, no."

"There was one test that Lance used to score particularly well in," said former teammate Frankie Andreu. "He produced less lactic acid than the rest of us." Ed Coyle also found this to be the case. "The most unique aspect of this individual's blood lactate profile," he wrote in the *Journal of Applied Physiology* study, "was the extremely low lactate concentration measured 4 min after exhaustion during measurement of VO_2 Max. Maximal blood lactate in the trained state was only 6.5–7.5mM in the present subject. By comparison, all the competitive cyclists we have tested, including team mates training with this subject, possessed maximal blood lactate postexercise in the range of 9–14mM" (mM denotes a concentration of one millimolar per liter).

Towns asked Ashenden if, in his opinion, Armstrong's blood lactate level was an explanation for his outstanding performance. Coyle's recorded lactate figures were shown on the screen again.

> **MA:** This is the value here that we've been talking about, maximal blood lactic acid, 7.5, 6.3, 6.5, 6.5, and 9.2. Another argument put forward to explain this dramatic increase in performance is the notion that the highest lactic acid levels that Armstrong produced were remarkably low. There's two problems with that argument. First of all, these values are low, but, again, you see that kind of value in a professional cyclist. Now, it's the bottom end, I acknowledge that. But it's not so low that you would say, well, there it is, the magic bullet. But perhaps the more important thing is that I, as an exercise physiologist, I would struggle to find a colleague that would put up their hand and say, yes, your maximal lactic concentration is able to predict your performance. It's been misconstrued, and I'm not sure why he [Coyle] has pushed this point so . . . so diligently. But the truth of the matter is that back, say, ten, fifteen, twenty years ago, lactic acid was viewed in a completely different context. It was thought of as this evil thing. You know, lactic acid impairs your muscle function, blah, blah, blah. The most recent literature turns out . . . lactic acid really is good during exercise. It's essential. It's used as a fuel by the muscle, so the notion that having a high level is going to give you, or a low level is going to give you better performance is flawed.

Ashenden later commented that the key point is that low levels of lactic acid were not, of themselves, going to lead to better performance, but if low levels of lactic acid are recorded using the same workloads at different times, that is a positive thing.

In explaining Armstrong's physiological advantages, Ed Coyle spoke glowingly about the rider's heart. Not only was it as big a heart as you would find in a seven-foot man, but the rate at which the heart beat was also extraordinary. "It's very rare," he said, "to see competitive cyclists and especially people his size with heart rates above 190. Certainly, very few. I've never seen anybody with heart rates above 200. His was 207. And you can see it remains high. He's human. It does come down with age, that's typical of everybody. We lose about one heartbeat per year at maximum. Still his maximum heart rate is 200 beats per minute. That gives him, in and of itself, a 5 to 10 percent advantage over other bicyclists. A heart that can beat more frequently at maximum while still pumping a lot of blood is going to have that advantage."

Nothing, it seemed, could have been clearer.

Ashenden was scathing in his dismissal of this argument. "The heart rate, it's been around physiology for a lot of years. I'm going to give them [Coyle and his associates] benefit of the doubt and suggest that they probably didn't understand what they were actually saying. It was a mistake. To suggest that someone's maximal heart rate gives them a performance advantage over someone whose maximal heart rate's low is nonsense. I can't understand why physiologists that have worked with the USOC [United States Olympic Committee]—and I think Eddie Coyle said he'd seen a thousand cyclists or something—why would they say that? I think they were probably mistaken, but you wouldn't make that assumption."

The Australian was equally skeptical about the contention that Armstrong's heart was unusually big. "Well, again, there's another explanation, at least put forward in the press, and I know I read it, and it's been repeated on that DVD that we saw from the Discovery Channel that, you know, Lance Armstrong's heart is an incredible size, and I think in the interviews, I saw Eddie Coyle say it's equivalent to the heart of a seven-foot man. I was skeptical about that from the outset. I've never seen any measurements, and it just seemed rather a convenient explanation. And it's simple enough to do, to measure heart size, and based on what we've heard, they've never measured it apart from an echocardiogram. . . . Looking at the [medical] reports leading up to his cancer treatments, they noted that his heart was within normal limits. Now, they said it was on the upward boundary of normal limits, but that, to me, is a much more objective way to look at it and it doesn't suggest that—according to the Armstrong stats, and I think Lance Armstrong said as much himself—that it's an incredibly large heart. I don't think Ed Coyle had any basis to suggest that it was equivalent to the size of a seven-foot man. I don't know where he would have made that assumption from."

The bottom line in Ed Coyle's analysis is that over a period of seven years (1992–1999) Armstrong achieved an 18 percent improvement. According to Coyle, this happened in two ways: an 8 percent gain in raw power from muscle adaptations brought on by intense training, and an 8 percent improvement from reducing his body weight. The coefficient of these two factors resulted in an overall 18 percent gain. "The simple point is that you can improve that much, you can win. You can win," Coyle told the court of arbitration. "And it just places these other things like r-EPO into some quantitative perspective." Coyle claimed the effect of r-EPO would be to improve power output by 5 to 6 percent, so it was entirely understandable that if Armstrong improved 18 percent, he was going to have no difficulty beating doped rivals.

The rhetoric was better than the science. No study was quoted to support the view that r-EPO generated an improvement in power output of 5 to 6 percent, and the means by which Coyle came up with his 18 percent improvement are questionable. In the calculations that produced this figure, Coyle used a figure of 72 kilograms (approx. 159 pounds) for Armstrong's weight, although in the five testing sessions at the University of Texas lab, 75.1 kilograms was the lowest recorded weight. Coyle said Armstrong reported it was 72 kilograms in the Tour de France, and in a scientific paper, he was happy to take the rider's word for it.

At one point in his evidence, Coyle was explaining the importance of Armstrong's reported weight loss when he unwittingly lapsed into the jargon of the lecturer. "And hopefully in a question and answer period I can follow up on the idea that Lance is literally—he goes hungrier, he's hungrier. He says for the two or three months before the Tour de France, 'I'm going to be hungry. I'm going to lose some body weight and that's important. I'll do what it takes because I know that saving ten pounds is remarkably important.' Why his European competitors don't do the same is just beyond me."

Coyle didn't explain how he knew European riders didn't attempt to reduce their body weight. Had he interviewed a cross section of European cyclists? Armstrong, in an interview with Charlie Rose on PBS television, aired in August 2004, admitted his percentage of body fat at the time of the Tour de France is not as low as some of his rivals.

Charlie Rose: You have no body fat?

LA: That's not true.

CR: You have a tiny bit?

LA: It depends when. I mean, during the month of July, I have the right amount. But I'm not the skinniest guy on the tour. There are guys who are 3 percent [body fat]. But I'm typically 5, 6 percent. But in the winter time, I get up to 10 or 11 percent.

In his article for the *Journal of Applied Physiology*, Coyle listed Armstrong's weight as he recorded it at their five testing sessions. They were:

November 1992—78.9 kilograms
January 1993—76.5 kilograms
September 1993—75.1 kilograms
August 1997—79.5 kilograms
November 1999—79.7 kilograms

Among the many interesting points about the figures, what strikes one immediately is the direct comparison between Armstrong's November 1992 weight and his weight in November 1999. In '92, three months after turning professional with the Motorola team, he was 78.9 kilograms; in '99, four months after winning his first Tour de France, he was 79.7 kilograms (a difference of less than two pounds, but Armstrong weighed *more* in 1999). This raises questions about the reliability of Armstrong's reported loss of weight and the part it played in his Tour de France victories. According to the consensus view of the time, cancer and the treatment that followed changed his body shape, and with greater discipline, he supposedly lost anything from fourteen to twenty pounds. There is no evidence to support a weight loss of this magnitude.

Perhaps the most interesting figure for weight offered by Professor Coyle is the 75.1-kilogram measurement taken in September 1993, around the time Armstrong won the world championship road race at age twenty-one. Coyle says this constituted his racing weight at that time, and there is no reason to question that. So if Armstrong dropped even 5 kilograms postcancer, his weight at the start of each Tour de France would have been 70 kilograms. Ed Coyle states Armstrong reported his weight as being between 72 and 74 kilograms during the Tour, and at another point, Coyle claimed it was 72 kilograms. Considering the question of his postcancer weight in the Tour de France, Armstrong's testimony to the tribunal was revealing: "I would love to have started every Tour at seventy-two or seventy-two and a half, that was the dream, and I never really did, perhaps, get down to that weight. But generally, if I saw the low seventy-fours, I would be happy."

According to tests done by Coyle, Armstrong's racing weight in 1993 was 75.1 kilograms. According to Armstrong's sworn testimony, he was happy if he started his postcancer Tours with a weight in the low seventy-fours. The conclusion is indisputable: postcancer, Armstrong was one kilogram (approx. two pounds) lighter than in the precancer years. While his body shape may have changed, there is no evidence to support the widespread belief that he achieved substantial weight loss.

During the French investigation into the U.S. Postal Service team after the 2000 Tour de France, fifteen urine samples given by Postal riders during the race were later examined by Dr. Gilbert Pepin, a Parisian-based toxicologist, and some time afterward by Professor Michel Audran from the Department of Pharmacology at the University of Montpellier. They found virtually nothing in the samples, thirteen of which belonged to Lance Armstrong. Rather than being reassured by this, they were suspicious. SCA lawyer Cody Towns asked Dr. Ashenden about the clarity of those samples.

> **CT:** What do the samples being too clear indicate to you?
>
> **MA:** Clear urine is inconsistent with an athlete who has just been on a bike for four, five, or six hours and raced up a mountain, and it is consistent with urine substitution, and that concerns me because the year 2000 was the time where there was an enormous amount of publicity about the work to develop a test for r-EPO. And I think that if you look at that and take into account the fact all of his urine samples were clear and it was during the mountain stages, as well, that causes me concern.

Ashenden also recalled instances from the 2004 Olympics in Athens in which medalists concealed quantities of bogus urine on their bodies and successfully defrauded doping control. The point was clear: if it was possible at the Olympics, where testing protocols are so stringent, it could happen anywhere.

Ashenden also looked at Armstrong's six alleged positive tests from the 1999 Tour de France, discovered in research testing six years after the fact, and tried to correlate the dates of the positive tests with the demands of the race on those particular days.

"Now, in examining these 1999 results," SCA's lawyer Towns asked him, "have you been able to do comparisons of those results with Mr. Armstrong's performance in the 1999 Tour de France?"

> **MA:** Yeah. I overlayed the results and the dates from the doping control forms with the stages that were raced in '99.
>
> **CT:** Can you tell us, when you did this comparison, what is it that you found?

Before matching Armstrong's six positives to the stages during which the samples were collected, Ashenden first explained the method by which a sample would be considered positive. If the level of basic isoforms exceeded 80 percent, this was deemed evidence of exogenous r-EPO and the test declared positive. Sometimes the level is in the 80s; on other occasions it is closer to 100 percent, depending upon the proximity of the test to the r-EPO injection. "To find 100 percent," said Ashenden, "is consistent with an injection that had been given a couple of hours before the sample had been collected." Ashenden thought it was significant that Armstrong's July 3 test found basic isoforms at 100 percent, indicating the injection of r-EPO was taken a few hours before the sample was collected. July 3 was the day of the 6.8-kilometer time trial and the shortness of the race made it possible that a rider had taken an r-EPO injection a few hours before the sample was collected.

> **CT:** Walk us through what you see in the stages after the prologue.
>
> **MA:** Stage nine, he's tested and you see 96.6 percent. Now, that's less than 100 percent obviously, and, again, it's consistent with an injection that he would have received, could have received earlier in the day. The next day they test the samples and there's 88.7 percent. Now, that's what we saw in our research when we tested an athlete every couple of hours. The percentages come down. . . . He was tested on the

twelfth stage, 95.2. . . . And then on the fourteenth stage, 89.4 percent. There was a rest day. From this point on, Armstrong had a four-minute, forty-four-second lead. . . . I find it unusual that from that point forward, there was never enough EPO in any of his urine samples to report a result.

CT: Why is that unusual?

MA: It's unusual because when an athlete stops taking EPO, it is no longer the injected EPO that gets into urine. Their own kidney has shut down production of EPO because the body recognizes that there's too much blood in his circulation. It suppresses EPO production so it gives your body a chance to come to its natural level. It's consistent with not finding enough EPO in the sample to analyze. You see that when an athlete stops taking EPO injections. And, again, this is where I'd suggest that if the laboratory was, for whatever reason, spiking samples, without knowing which samples corresponded to which day, the fact that there's this consistent pattern before and after, to me it's inconceivable that it could be a result of deliberate tampering.

CT: Now, Dr. Ashenden, I want to ask you, looking at all of the evidence that you've seen, in your own experience as a scientist and expert in this field, do you reach a conclusion as to whether Mr. Armstrong has used performance-enhancing drugs in his career?

MA: I think that as a physiologist, I look at the unexplained jump in performance. As an anti-doping researcher, I look at the strange changes in the blood. As a layperson, I look at the admissions that he admitted to using these banned drugs and that that would have explained this previously unexplained jump in performance. I bring into the equation that you've analyzed the samples, and it shows that he was using EPO during the '99 Tour de France. I have to conclude that beyond any reasonable doubt, he has used performance-enhancing drugs.

There is an apparent contradiction in Ashenden's argument when he says Armstrong's alleged admission of drug taking, made in late October 1996, would have explained his sudden jump in performance from 1999 onward. If Armstrong was using performance-enhancing drugs in 1996 and presumably before, why did the greatest improvement take place in 1999? Michele Ferrari is, perhaps, the key to understanding the rider's progress. After becoming the sport's youngest ever world road champion in 1993, Armstrong endured two relatively disappointing years in 1994 and '95. In '94, he didn't win one race in Europe. At the end of 1995, he began working with Ferrari and, according to his teammate Frankie Andreu, he was a different rider in the spring of 1996. Andreu says he was more muscular and far more powerful than before, and Armstrong's results did improve significantly through the first four months of '96, before the onset of testicular cancer temporarily halted his career.

Professor Ed Coyle's thesis is that from age twenty-one to twenty-eight, Lance Armstrong improved steadily from being a rider with a lot of potential to one who was the greatest performer in the history of the Tour de France. The improvement was, according to Coyle, constant and consistent. He described the graph charting the progress as being a straight line upward. Given that Armstrong was struck down with cancer in 1996 and did not race in 1997, much of the gradual improvement would have had to take place in the years 1993 to 1995.

To win the Tour de France more than once, the would-be champion has to be an outstanding time trialist. Coming as it usually does at the end of the first week or at the beginning of the second week, the long individual time trial separates the contenders from those who aren't good enough. For this reason, it is often called the race of truth. Lose three or four minutes in the time trial and more often than not, the Tour is lost. In his first three Tours, Armstrong was remarkably consistent in his race against the clock: at the Lac de Madine time trial in 1993, he finished twenty-seventh, 6.03 minutes down at the end of 59 kilometers. A year later he finished 6.23 minutes down over the slightly longer 64-kilometer route, and in 1995, when the time trial came in the final week, he was 6.24 minutes down at the end of 46.5 kilometers.

Which is certainly not to imply that Armstrong took it easy in those time trials—his natural competitiveness would not allow that. Interviewed after the 1993 effort, he said: "The time trial, it's the biggest thing on my mind. I know I gotta learn how to do it. At the same time I know I can do it. I lost six minutes. I mean, it's as plain as day, I know exactly where I am. I went all out, Indurain went all out. I'm six minutes down. But he's the greatest time trialist there's ever been. If I can get a minute a year, a minute a year isn't that much. I'm twenty-one, he's twenty-nine. When you're thirty you're not gonna be nine minutes faster than you are at twenty-one, but maybe in three years I will be three minutes faster."

In the three years to which Armstrong is referring, Miguel Indurain won the long time trial and offered the same standard by which others could measure their progress. Armstrong's deficits were 6.03, 6.23, and 6.24. He was what he was in time trials: solid, consistent, but nowhere near the best, and over three years, he made no improvement. In 1996, Armstrong, feeling sick, abandoned the Tour toward the end of the first week and didn't get to ride the individual time trial. He would next ride the Tour de France in 1999 and destroy every rival in the 59-kilometer race against the clock at Metz. A year later he would again blitz his rivals in the time trial. His average speed in that 58-kilometer test at Mulhouse, 53.98 kilometers per hour, was the fastest for a time trial over 30 kilometers in the history of the Tour de France. And the man described by Armstrong as "the greatest time trialist there's ever been," Miguel Indurain, had never gone so fast.

In the 2003 Tour de France, three years after Armstrong's record, the specialist in the time trial, Scotland's David Millar, went marginally quicker over a route six kilometers shorter. A year later, Millar admitted to French police that he'd used EPO to enhance his performances in 2003.

Chapter 20

"I'LL SAY NO"

It wasn't supposed to be like this.

Stuck in a hotel room in Paris, yesterday's winner telling today's fibs. Four days before, Floyd Landis had stood on the Avenue des Champs-Elysées and felt like he owned it. There are yellow jerseys and there is *le maillot jaune* presented on the final day of the Tour de France. With that, you've got it made. And what made it special for Landis was the presence of his wife, Amber, and his stepdaughter, Ryan. The millions of dollars could wait, so, too, the talk shows, the fame, the endorsements, because nothing would transcend this moment. And how quickly it passed. One phone call from the Union Cycliste Internationale (UCI) and his world was shattered. From the most celebrated cyclist to the most notorious cheat. Who prepares you for this? What to do when you can't say, "Well, things could be worse"? Never in the 103 years of the Tour de France had the winner been officially called a doper, much less just four days after the victory. Not certain what to do, those close to Landis told him to fight back and that meant responding to the allegation. Even if it's true that the managers, agents, trainers, and PR flunkies need the star to remain in the firmament in order to maintain their careers, they are also fans—especially in Landis's case. If you knew him, you liked him. Tell 'em it ain't so, they said. And so it was that four days after his triumph, the winner of the 2006 Tour de France was holed up in a Parisian hotel defending his reputation.

This teleconference happened before Landis had worked out a strategy. It was like an NFL team lining up to receive the ball with no idea what they were going to do with it. For Landis, this was particularly difficult because it is not his nature to duck questions or tell cock-and-bull stories. This eldest boy of plain-speaking Mennonite people was not one for PR games. He is what he is—a slightly awkward but wholly sincere young man from a deeply conservative and religious community in the rural outpost of Farmersville, Pennsylvania. "Probably the least hypocritical professional cyclist I've met," Jonathan Vaughters said in an interview with a journalist from *The New York Times,* and that made Landis the least suitable candidate for the interview he faced that Thursday in Paris. He was no more convincing the following day when conducting a similar interview in Madrid. Landis's difficulties in those interviews stemmed from his immediate and uncomplicated view of the situation: he believed he had to explain how his urine sample from stage seventeen of the 2006 Tour de France contained a testosterone–to–epi-testosterone ratio that was almost three times greater than the allowed 4 to 1 ratio. Later, he would learn more clever ways of tackling the problem.

But, in those first two days, he clutched at every straw the wind blew his way. The high level of testosterone could have been caused by the beer and/or the shots of Jack Daniel's he drank the night before the stage. *How much alcohol did you consume?* "I didn't sit down and drink until I passed out," he said. *So exactly how much?* "I don't remember." At the Madrid press conference the following day, he was reported as saying it was "two beers and at least four shots of

Jack Daniel's." He spoke without conviction; his halting voice betrayed uncertainty, the explanations lacked logic. Dehydration was mentioned as a possible cause of his elevated T:E ratio, but anyone watching stage seventeen saw how much water Landis drank and poured over his head. His trainer, Allen Lim, quantified the rider's water use on stage seventeen in a diary he wrote for *Bicycling* magazine's website: "Taking, no joke, a total of 70 water bottles (480 ml each) from the car to keep himself cool and hydrated." Even on the hottest days, it's hard to get dehydrated on thirty-three liters of water.

Then there was his "naturally high level of testosterone," which was akin to Tyler Hamilton telling the Phonak doctor, Iñaki Arratibel, that his positive test for a mixed blood cell population was caused by a surgical procedure and transfusion he'd undergone sometime before. If true, it was the perfect alibi. But it was also an explanation that could be checked out. Hamilton quickly realized this and never again mentioned his "surgical procedure." When journalists heard the possibility of Landis's naturally high levels of testosterone, they inquired about other drug tests he'd had done. Could he produce evidence that showed his naturally high levels? And so that explanation perished. There were suggestions, too, that the cortisone he was legally allowed for his troublesome hip had somehow affected his T:E ratio. Not likely, said the experts.

Another suggestion was the high number of thyroid-stimulating hormones that Landis naturally produces, causing him to have Hashimoto's disease. "I don't know exactly how it works," he said of this possible explanation, "but, anyway, your thyroid does not function properly and you need to take, each day, a thyroid pill—a small minor inconvenience. Anyway, I don't know if it has any implication at all on what is going on here." His doctor Brent Kay was asked if the thyroid medication could affect the T:E ratio. "We are consulting with a number of the world's experts so that we don't speculate at this point." But if the cortisone or the thyroid medication affected Landis's T:E ratio, why hadn't it shown up before?

So the Paris and Madrid teleconferences rolled on. Though it was stressful for Landis, it could have been a lot worse. Journalists' questions at such times, especially when asked by sportswriters, tend to be respectful when not downright apologetic. And for understandable reasons: few cycling writers want to incur the wrath of a subject whose cooperation they will need in the future. In interviews with Daniel Coyle for his book *Lance Armstrong's War*, members of the U.S. Postal Service team told the author how the team secretly photographed journalists—"For the black list," spokesman Jogi Muller joked, but Coyle did not believe it was a joke. "We have friends in the pressroom," directeur sportif Johan Bruyneel said. "We know what goes on, and who is with who." In fact, through the last two years of Armstrong's career and even on his return to the Tour de France as the retired champion, entry to his press conferences was by invitation. Only those journalists considered supportive of—or, at the very least, sympathetic to—the team were asked along. Perhaps as significant, many journalists empathize with the lie. A regular Tour de France writer knows doping has been a central part of the race's history for decades. Yet it is tough to prove and anyone spending too much time on the doping question risks damage to their enthusiasm for the sport, not to mention the smearing of their reputation in the peloton. Paul Kimmage, the former professional cyclist who is now an outstanding sportswriter, tried to set up an interview with the Scottish cyclist David Millar in May 2004 but was turned down. "You've got a bit of a reputation," Millar told him. Kimmage's crime had been to write the story of his life as a pro in the seminal anti-doping book *Rough Ride*. The central point of *Rough Ride* was that the system virtually forced riders to cheat. But Millar was right; the book earned Kimmage a bad reputation among his former colleagues. (Two months after Millar turned down Kimmage, he was persuaded to do another kind of interview with French police. It lasted five days, and at the end of his incarceration, Millar admitted his use of r-EPO.) Many good journalists have left cycling, preferring to work in places where the truth is more accessible. And others still working in the sport accept the lie, and in many cases they empathize with it. Without the empathy, they could hardly do their jobs.

During the Friday teleconference in Madrid, Landis was asked if he intended to carry out further tests to determine if his body produced the naturally high levels of testosterone he had spoken about. "I understand," he said, "there is a process by which we can take a sequence of tests to prove there is a variation or there isn't a variation in the T:E level." The follow-up question asked if he would do the carbon isotope test that distinguished exogenous (synthetic) testosterone from natural testosterone. "I don't know if that would actually help my case. What I've been accused of is a ratio that is higher than normal. Maybe you're right, I don't know." Everyone remained polite, because there is no fun observing an intelligent man come up with stupid answers. No one screamed, "Look, Floyd, if you hadn't used synthetic testosterone you would be desperate to get the carbon isotope test done, as it would prove the inflated T:E ratio was not caused by synthetic testosterone." Instead the questioner patiently pointed out it would bolster his case if it could be shown there was no synthetic testosterone in his sample. Faced with the undeniable logic of this, Landis said, "I will keep that in mind."

Among those asking questions of Landis was Juliet Macur of *The New York Times.* As one who hadn't spent time in Europe waiting for the big players in professional cycling to give five or ten minutes of their time, and as one who hadn't had to walk the ethical tightrope that comes with a working life in the sport, Macur felt free to ask whatever questions she wished. "Let me get this straight," she said to Landis during the Paris teleconference. "You are saying that you have never taken any performance-enhancing drugs or methods during this Tour or beforehand. That this is a total mistake?" The question could hardly have been clearer: *Have you doped, ever?* "I am saying," Landis replied, "that I don't know what the explanation is for it [the high T:E ratio]. Whether it is a mistake or whether it is an occurrence from other circumstances that go on during the race or something that I did. The explanation I am saying is that it is not from an exogenous outside source of testosterone. That is what I am saying."

Clearly feeling the question had not been answered, Macur persisted. "Have you ever taken performance-enhancing drugs before?"

Confronted by an even plainer version of the same query, Landis paused. "I'll say no," he replied. When the question is framed in absolute terms, it invites an unequivocal response. "I'll say no," was startlingly equivocal. Landis tried to explain the reason for his qualified denial. "The problem I have is that most of the public has an idea about cycling because of the way things have gone in the past, so I will say 'no' knowing that a lot of people are going to say that I am guilty before I have had a chance to defend myself." Landis reasoned that because people would presume he was guilty, he decided to begin his defense with a qualified denial of the accusation. Those journalists who wanted to believe Landis found "I'll say no" demoralizing.

At the following day's teleconference, Philip Hersh of the *Chicago Tribune* asked Landis what he had meant by it. "I was a little bit overwhelmed," he replied. "I was trying to be fair and I was trying to picture what it would be like to be someone else looking at the situation. And so what I was saying, in a roundabout way, which should have been more direct, was that I'll say no, but I am saying it at the risk and knowing that probably a lot of people are thinking, 'Well, here we go again, this is the same old story.' So what I was trying to ask for at the time was that people, for just one more time, presume that I am innocent until I have a chance to defend myself. It took a long time to say it. I should have said, 'No, I'm not guilty.' That's what I meant to say. It came across as a little bit drawn out, and I apologize for that."

To allow the champion to regain his composure, easy questions are offered. Landis was asked if he'd had support from his former teammates on the U.S. Postal Service and Discovery team. There was, he said, a text message from George Hincapie and a telephone call from Lance Armstrong. He was asked what Armstrong had said. "He told me, 'Just tell them the truth, tell them you're not guilty, keep it short so they don't get confused. I've been through it, and you'll be all right.' " The Madrid teleconference ended when Landis realized he was in danger of missing his next appointment. "Sorry, guys," he said. "I got to get out of here. I've promised Larry King I would talk to him."

His performance improved with practice. The more interviews he did, the less awkward he seemed. He benefited from an important change in strategy. Rather than try to explain his 11 to 1 testosterone–epi-testosterone ratio and the discovery of synthetic testosterone, he focused attention on the French lab and the protocols of drug testing. The best defense was a good offense and he acted on advice from his friend Tyler Hamilton to make public as many details of the case as he could. Using the Internet, Landis published the 370 pages of laboratory documentation provided to him. Accompanying the documentation was a PowerPoint presentation attacking the credibility of the results. It was impressive and as the French laboratory was guilty of some sloppy administration, Landis and his team had something to work on. An important member of that team was Dr. Arnie Baker. They first met in 1997 and soon became friends. When the now retired Dr. Baker met Landis at the funeral of Landis's close friend and father-in-law, David Witt, a month after the 2006 Tour de France, they talked about his case and Baker offered to help Landis put his defense before the public. His presentation expressed damning opinions about the work of the French lab.

He questioned whether the sample that was purported to be Landis's was in fact his, and he used laboratory errors to support his argument. The identification number for Landis's sample was 995474, but on the page summarizing the A sample results it is typed as 994474. "It's an error as regards numbering," said lab director Jacques de Ceaurriz, "a typing error which has no significance whatsoever on the findings in the sample. These little mistakes happen. They are corrected, and noted." Embarrassingly for the lab, that wasn't the only administrative error. On another page summarizing the A sample results, Landis's identification number has been overwritten; correction fluid is used to obliterate what was previously written. Good laboratory practice calls for a line to be drawn through the incorrectly typed or written information and for the correction to be made alongside it, after which it is dated and initialed by the person making the change. In chain-of-custody documentation for the critical stage seventeen of the Tour, the handwriting for Landis's identification number is ambiguous and 995474 could be mistaken for 995476. These are important issues for the UCI and the French laboratory, and Baker concludes that the sample may not be Landis's. This could be resolved by DNA testing.

When the sample from stage seventeen was initially screened, it revealed a T:E ratio of 5.1 to 1, marginally above the permitted 4 to 1 threshold. Drug testing protocol demands further investigation of any sample with a greater than 4 to 1 ratio, and in most cases the result is that the initial finding is not confirmed. Landis's subsequent screening not only confirmed the abnormal ratio but revealed the ratio to be far higher than the initial screen suggested. Baker questioned the 11 to 1 finding, claiming the results showed evidence of contamination. He based his argument on a WADA stipulation that "the concentration of free testosterone and/or epitestosterone in the specimen is not to exceed 5 percent of the respective glucuroconjugates." Anything higher than 5 percent is an indication of possible contamination and Baker argued the percentage of free epitestosterone was 7.7 percent. There are other indicators of contamination—such as the pH of a urine sample—that he does not mention. Landis's sample did not have an abnormal pH.

If Landis's sample had been contaminated, that would still not explain the presence of synthetic testosterone. How did it get there? He realized this was the key accusation and during his initial interviews, before the existence of synthetic testosterone had been publicly revealed, he spoke of the laboratory finding being nothing more or less than an unusual T:E ratio. He knew what he was doing because there are many ways to explain an elevated ratio but precious few satisfactory explanations for synthetic testosterone. Landis's team focused on the protocols of the carbon isotope test, especially the phrasing of guidelines in relation to the examination of the breakdown products, or metabolites, of testosterone. The French lab works on the basis that if one of the metabolites is abnormal, that is evidence of synthetic testosterone. Howard Jacobs, Landis's lawyer, seized on the use of the word "metabolite(s)" in the WADA protocols and claimed it wasn't clear whether all metabolites must be abnormal or just one. Jacobs pointed to the IOC-approved lab at UCLA, under Dr. Don Catlin, which checks for two metabolites and needs both to be abnormal before declaring the presence of synthetic testosterone.

Dr. Christiane Ayotte, head of Canada's national anti-doping laboratory in Montreal, told a journalist for cyclingnews.com that any one metabolite showing a significant difference from the negative control indicates a positive test. Speaking of the drafting of the WADA document and the use of the word "metabolite(s)," she said the intention was to state that if one or more metabolites was abnormal, that confirmed the existence of exogenous testosterone. "We never imagined that this would be taken in any other way," she said.

In the promotion of his campaign, Landis has shown the same doggedness and focus that marked his career as a racer. He was also backed by a strong team and was especially indebted to Arnie Baker, Howard Jacobs, and Michael Henson. Dr. Baker looked after the science, Jacobs took care of the legal aspects of the case, and Henson organized the PR. No one played a more active role than the rider himself, and after setting up the Floyd Fairness Fund, Landis embarked on a series of town-hall-style meetings in the United States to drum up support for his case and raise funds to help defray the cost of his defense.

In one of the slides prepared by Baker, it was argued that Landis's positive test didn't make sense because the "samples before and after were okay." This was interesting because it indicated Landis's team was ready to accept the French lab's work on the rider's seven other samples from the 2006 Tour de France. USADA's general counsel Travis Tygart was also interested in these samples, but for a different reason. In late December 2006, he asked permission to do the carbon isotope test on the seven negative samples to see if any contained synthetic testosterone. Because the T:E was less than 4 to 1 when these samples were originally examined, the carbon isotope test would not have been carried out. Tygart's request was an opportunity for Landis to use the remaining samples to help show his stage-seventeen positive was the result of one test being botched. Landis refused the request, citing his lack of confidence in the French laboratory. He also questioned why USADA would make such a request, "unless it's simply another way to drain my resources in this fight to clear my name." In fact, it was easy to understand USADA's request because if the other samples contained synthetic testosterone, it would greatly undermine Landis's case.

The argument continued over whether it was permissible to examine B samples of urine that had already been tested and cleared. Landis's team claimed it would be illegal, but the WADA legal director, Olivier Niggli, disagreed. "The World Anti-Doping Code allows additional analysis on B samples when there is a need for it in the procedure." USADA pressed for the additional tests, and in the third week of April 2007 Landis's seven other samples were tested for synthetic testosterone. Several of them showed evidence of synthetic testosterone. Landis's team claimed its representatives were not allowed to witness the tests taking place and accused the authorities of malicious behavior. USADA and the French laboratory are not allowed to speak of individual cases and their response to the allegation of maliciousness would not be heard until Landis's formal hearing took place. A source close to the French laboratory claimed that both USADA and Landis's representatives had been present through the week, but when the testing ran over into Sunday, USADA's people could not remain in Paris for the extra day. According to the source, the agreement was that both sides' representatives could attend the testing, but if one side could not be present for a session, the other side was precluded from attending that session.

Not everyone was impressed by the Landis campaign. Christophe Moreau, the French rider who had once been part of the Festina scandal and now rides with the AG2R Prévoyance team, spoke to a journalist from *L'Equipe*. "Here, we have reached rock bottom. He

[Landis] has created a fund to gather money for his defense. Landis is positive with his A and B samples and he has the nerve to go begging for dollars while his lawyers look for the comma that is out of place, that will show up some technical irregularity."

The sadness of the story lay in the violation of the order of sports. There is a race; it lasts more than three weeks; we follow its course, waiting to find the winner; but when it ends, there is no victor. Three, six, nine months after the end of the Tour, there is still no one whom we can call the 2006 champion, and it gets to a point where it doesn't much matter anymore. Soon after the announcement of Landis's positive test, the runner-up, Oscar Pereiro, was asked how he felt about the likelihood of eventually being promoted to the top position. He said it would be like an "academic victory," suggesting he didn't have much enthusiasm for the disqualification of Landis and wouldn't take much joy in triumph by default. Over the weeks and months that followed, Pereiro changed his attitude and seemed more prepared to accept victory and the winner's prize money, however it came about, but that initial reaction was not unusual in cycling. When it looked like the Spaniard Pedro Delgado would lose the yellow jersey because of a doping violation in the final week of the 1988 Tour, the second-place rider, Steven Rooks, was asked how he felt about winning the race in this way; the Dutch rider said he didn't want it.

"But if Delgado has cheated, you will deserve to be the champion."

"Pedro," he said, "has been the strongest rider. It is he who deserves it."

Delgado eventually escaped on a technicality. Probenecid, the masking agent found in his urine and on the IOC's banned list, would not be added to the UCI list until a week after the race ended. Twelve years later, Rooks admitted to using testosterone throughout his career and confirmed the impression conveyed in his reluctance to see Delgado as a cheat.

If the scandal that came at the end of the race poisons memories of the 2006 Tour de France, it also destroyed one of sports' more romantic victories. Landis's story should have lifted our hearts, not broken them. His should have been the classic story of rebellion and resilience, reconciliation and resolution. He challenged the strict Mennonite way of life and when he told his story of how he and his well-intentioned parents fought, he recalled the ageless family struggle.

Landis rode at night because Paul, his father, filled his days with so many chores there wouldn't be time to waste on a racing bike. Not to be bested, the boy rigged his bike up with a headlamp, put on extra clothing to combat the icily cold weather, and set himself a target of five hundred miles of training every week. At the age of twenty, he decided to leave home and chose California to put as much distance between himself and the world his parents had chosen. As he said in pretty much every interview he did, it wasn't that he didn't love his parents, but that he didn't want the Mennonite way of life. He needed to go, as much to assert his independence as to create a different life for himself. Since the days of his lonesome training rides in Lancaster County, he had sensed he was good enough to be a racer.

He started as a mountain biker, but as the craze for that branch of the sport waned, he switched to road racing, where the top three or four hundred racers could make a decent living. Mercury-Viatel was his first road team and he soon turned heads with his impressive third-place finish in the French Tour de l'Avenir. After spending three years with Mercury, he joined the U.S. Postal Service team in 2002, and in three years with the team, he proved to be a high-class équipier for Lance Armstrong. Landis's value to Armstrong was seen on the critical mountain stages, when he would selflessly help set the pace with the team's other climbers, so that Armstrong got the smoothest ride possible until it was time for him to make his move. From 2002 through 2004, Landis and George Hincapie were the only Americans in the Postal squad that Armstrong considered strong enough for the Tour de France.

Through the first two seasons, Landis fitted easily into the team and seemed to enjoy a reasonable rapport with the team leader. Before the first two Tours with Postal, 2002 and 2003, Landis spent five weeks with Armstrong and Michele Ferrari close to the doctor's office in Saint Moritz in Switzerland. Despite the doctor's tarnished reputation, Landis didn't express any reservations about him. Not so with Armstrong, and their relationship would have its frosty and fractious moments, especially through their final year (2004). Armstrong's single-minded and monklike focus in the months leading into the Tour made life difficult for his teammates, but most of them were expert at tiptoeing around the leader without ever treading on his toes. Landis couldn't always be relied upon to move so nimbly, and in any case, he was done with genuflection the day he left the Mennonite fold. One warmed to his character—the stubbornness, the refusal to turn his backside and be branded like every other Postal rider—and the quirks of behavior that made him appear almost wacky. Though he had left the Mennonite community, he brought with him the humility of those who settle for a simple way of life. In California, he met and later married Amber Basile and became stepfather to her daughter, Ryan; they set up home in modest Murrieta.

He left the U.S. Postal Service team at the end of 2004, perhaps tired of Armstrong's relentless control but also excited by the improved salary and the greater opportunities offered by the Phonak team. How can a rider ever know what he is capable of if he spends his career in the service of another? Landis had to find out what he could achieve, and as had been the case for Kevin Livingston, Tyler Hamilton, and Roberto Heras, the only way to do that was to get out from under Armstrong's wing. He finished ninth in the 2005 Tour de France without ever landing a blow on his former leader, who was winning the race for a seventh consecutive time. He had expected more of himself, and although he wouldn't make excuses, Landis felt he could and would do a lot better in the 2006 race.

Landis probably didn't know a lot about Lance Armstrong's litigation with SCA Promotions, which began in late 2004, carried on through 2005, and did not end until early in 2006. In the heat of the legal battle, both sides attempted to secure every document that might help its case. Armstrong's team forced SCA to disclose an e-mail that showed that one of its in-house lawyers had rifled through a trash can for a piece of chewing gum discarded by the Tour de France champion. The lawyer, Chris Compton, wanted to see if Armstrong's DNA could be taken from the gum and compared with the DNA of the six positive samples from the 1999 Tour de France. SCA also knew an important instant-message conversation had taken place between Frankie Andreu and Jonathan Vaughters and it issued a subpoena to secure it.

The following is the complete transcript of the conversation between Vaughters and Andreu that took place on Thursday, July 26, 2005, four days after Armstrong had won his seventh consecutive Tour de France. It was six A.M. in Andreu's home state, Michigan; four A.M. in Vaughters's home city, Denver.

Cyclevaughters: frankie—hey—thanks for talking the other day

FDREU: no problem, where are you

Cyclevaughters: back in CO

FDREU: nice, I just got home, isn't it like 5am

Cyclevaughters: sometimes i think i'm going to go nuts
yeah
it's 5am

FDREU: I agree, I came home and the air conditioning is broken

Cyclevaughters: ouch

FDREU: did your kid grow twice it's size in the two weeks you were gone

Cyclevaughters: yeah, his feet look bigger for some [r]eason

FDREU: funny

Cyclevaughters: anyhow, i never can quite figure out why i don't just play along with the lance crowd—i mean shit it would make my life easier, eh? it's not like i never played with hotsauce, eh?

FDREU: I know, but in the end i don't think it comes back to bite you
I play along, my wife does not, and Lance hates us both
it's a no win situation, you know how he is. Once you leave the team or do something wrong you forever banned

Cyclevaughters: i suppose—you know he tried to hire me back in 2001 . . . he was nice to me . . . i just couldn't deal with that whole world

FDREU: I did not know that
look at why everyone leaves, it's way too controlling

Cyclevaughters: once I went to CA and saw that now [not] all the teams got 25 injections every day, i felt really guilty
hell, CA was ZERO

FDREU: you mean all the riders

Cyclevaughters: Credit Agricole

FDREU: it's crazy

Cyclevaughters: So, I realized lance was full of shit when he'd say everyone was doing it.

FDREU: You may read stuff i say to radio or press, praising the Tour and lance but it's just playing the game

Cyclevaughters: believe me, as carzy as it sounds—Moreau was on nothing. Hct of 39%

FDREU: when in 2000–2001

Cyclevaughters: so, that's when you start thinking . . . hell, Kevin was telling me that after 2000 Ullrich never raced over 42%—yeah moreau in 2000–2001
Anyhow—whtever

FDREU: After 1999, you know many things changed. lance did not
I believe that's part of whey kevin left, he was tired of the stuff

Cyclevaughters: funniest thing i ever heard—Johan and Lance dumped Floyd's rest day blood refill down the toilet in front of him in last yrs tour to make him ride bad

FDREU: holy shit, I never heard that. that's craz! ! !

Cyclevaughters: that's from floyd
he rode this year with no extra blood

FDREU: I never knew
he did great also

Cyclevaughters: yeah, i could explain the whole way lance dupes everyone

FDREU: what abut GH climbing the mountains better than azevado and the entire group

Cyclevaughters: from how floyd described it, i know exactly the metho[d]

FDREU: explain that, classics to climber
when did you talk with floyd

Cyclevaughters: i don't know—i want to trust George
but the thing is on that team, you think it's

normal
or at least i did

FDREU: i guess. anything with blood is not normal

Cyclevaughters: yeah, its very complex how [to] avoid all the controls now, but it's not any new drug or anything, just the resources and planning to pull of a well devised plan
it's why they all got dropped on stage 9—no refill yet—then on the rest day—boom 800ml of packed cells

FDREU: they have it mastered. good point

Cyclevaughters: they draw the blood right after the dauphine

FDREU: how do they sneak it in, or keep it until needed
i'm sure it's not with the truck in the frig

Cyclevaughters: motorcycle— refrigerated panniers
on the rest day
floyd has a photo of the thing

FDREU: crazy! it's just keep going to new levels

Cyclevaughters: yeah, it's complicated, but with enough money you can do it

FDREU: they have enough money. Floyd was so pissed at them this entire tour

Cyclevaughters: anyhow—i just feel sorry for floyd and some of the other guys
why would lance keep doing the shit when he clearly has nothing to prove—it's weird

FDREU: I know. me to. they all get ripped into for no reason
he's done now, thank god. but they will prove next year for Johan's sake that they are the greatest

Cyclevaughters: and then lance says "this guy and that guy are pussies"

FDREU: they won't stop
i agree

Cyclevaughters: then i've got tiger as one of my sponsors, and he loves to pick my mind . . . what do I say?

FDREU: You play dumb.
You can't talk with them about this stuff
I would think they would freak

Cyclevaughters: yeah, that's tough—i do, but it's tough
maybe they should freak
what about dan osipow or louise? do they know what's up?

FDREU: I know, I get tired of hearing how great Lance is, what a super person, etc. it's crazy and it's hard to not just tall people he is a cheat and asshole
I think not. they just run the team. They are never allowed in a hotel room or bus or anything

Cyclevaughters: every other team in the tour you could just walk right on to their bus and say hi
but disco won't let dan on?
all right

FDREU: my kids are waking up, I gotta run. Let's talk some more later

Cyclevaughters: i've had enough—i'm glad to be home—hopefully this won't affect my team or my kids

FDREU: I agree

Cyclevaughters: that's all i care about

FDREU: ciao

Cyclevaughters: see ya

This is a remarkable conversation because it involves two men who are former professional cyclists, former teammates of Lance Armstrong, and highly respected within the cycling community. They are also two intelligent men who have been more honest than most of their peers about the existence of a pervasive culture of doping within their sport. Yet they have also been restrained in what they have said—they don't accuse other riders of doping, and in their public utterances, they err on the side of caution. But in this instant-message exchange, information and opinions are offered without the customary restraints and qualifications. The authenticity comes from the intimacy of the conversation; these are two friends who once soldiered together on the U.S. Postal Service team and whose lives still revolve around cycling. They know people, they talk to people, and so they hear things not meant for public consumption. The story of the conversation is worth explaining.

Perhaps the first thing worth noting is the time of the conversation: six A.M. in Andreu's Michigan, four A.M. in Vaughters's Colorado (even if he, mistakenly, says it's five A.M.). They have both just arrived back to the United States from the 2005 Tour de France. After catching up, Vaughters gently chides himself for not playing along with "the lance crowd," saying his life would be easier if he did and then acknowledging that he didn't have the right to present himself as a paragon, "it's not like i never played with hotsauce, eh?" There is only one plausible explanation for what Vaughters is saying here: he doesn't understand why he doesn't just accept the attitude toward doping on Armstrong's team; it would save him a lot of hassle, and anyway, it wasn't as if he'd never used performance-enhancing drugs. He would later confirm this impression in an interview with *The New York Times*. "I don't have a halo over my head; I made some mistakes when I was a rider," he told Juliet Macur. The evidence indicates Vaughters was a predominantly clean racer, but by his own tacit admission, he did dope at one point. That is most likely to have happened in the lead-up to the 1999 Tour de France, which is the one spike in his modest but consistent career graph, when Vaughters won the Route du Sud and finished second in the Critérium du Dauphiné Libéré. (A Postal rider, who did not want to be named, complained that Vaughters had "done something wrong" and then been selected before him for the '99 Tour. The same Postal rider also claimed he saw nothing wrong with using r-EPO if, through racing or training, his hematocrit was beaten down from a natural forty-six to an unnatural forty-two.)

Vaughters and Andreu then complain about the environment around the U.S. Postal Service team, something Vaughters didn't want to deal with when Armstrong invited him to rejoin in 2001. For Andreu, it was much too controlling. When Vaughters left the team at the end of 1999, it was to join the French squad Crédit Agricole, and he describes his sense of guilt on discovering a team with a very different ethos from the U.S. Postal Service's: "once I went to CA and saw that now all the teams got 25 injections every day, I felt really guilty." There is an important typing error in that sentence because what Vaughters means to say is that "not all the teams got 25 injections every day" and seeing Crédit Agricole run a team that had zero injections made him realize "lance was full of shit when he'd say everyone was doing it." Andreu then complains that after 1999, a lot of things changed for the better but "lance did not," and that may have been part of the reason why Kevin Livingston left the team: he was tired of the doping culture.

There is then the most startling segment of the conversation, as it connects Landis, Armstrong, and Johan Bruyneel to blood doping. Vaughters tells a story from the 2004 Tour de France about Landis's "rest day blood refill" being dumped by Armstrong and Bruyneel, "to make him ride bad." When Andreu expresses his amazement at this information, Vaughters says he got the story from Landis. Given that all three characters in this incident were part of the same Postal team, it is reasonable to wonder why Armstrong and Bruyneel would not have wanted Landis to do well in the final week. One insider, who did not want to be named, said that the animosity was caused by Landis's decision to leave Postal for the Phonak team.

Vaughters then says he can tell how Armstrong dupes people because Landis has explained it to him. According to Vaughters, the team was involved in an elaborate and well-organized system of doping involving blood transfusions. He says the reason the team did not perform well on stage nine of the 2005 Tour was because their blood refills didn't arrive until the following day, a rest day. Vaughters explained how the cheating was organized: blood was drawn from the riders immediately after the Critérium du Dauphiné Libéré, three weeks before the start of the race. The body naturally replaced the extracted blood, and the reinfusions during the following month's Tour de France provided the riders with an enormous boost. Andreu wants to know how the team got the blood to the hotels during the Tour. Motorcycles, says Vaughters, with refrigerated panniers. And he adds that Landis, who was riding for Phonak while all of this was happening at the 2005 race, has a photograph that shows one of the motorcycles.

After all Armstrong has achieved, Vaughters wonders why he would "keep doing the shit," while Andreu is pleased that the seven-time champion has retired. One is then offered an insight into why cycling's secret has remained intact for so long. Vaughters asks what he should tell one of his sponsors who is curious about how things truly are in the sport. Andreu advises him to play dumb and to keep his sponsor. Vaughters wonders how much Dan Osipow and Louise Donald, two Discovery team employees who have been with the team almost from the beginning, know. Andreu reckons they know little because they are not allowed on the team bus, nor into hotel rooms. Every other team, says Vaughters, you could walk right onto their bus. But Discovery, referred to as "disco," will not allow its general manager, Dan Osipow, onto the bus. (Osipow left the team at the end of the 2005 season.) Andreu has then got to go, his three children are waking up.

When this instant-message exchange was made available to the Armstrong-SCA proceedings, the part about Landis's blood being dumped during the 2004 Tour was omitted because it was not strictly relevant to the case. It was embarrassing for both Andreu and Vaughters to have so private a conversation seen by third parties. Although the SCA arbitration was confidential, much of what was said in depositions and then over the three weeks of the tribunal was leaked into the public domain. Both sides made available documents and information that they felt helped their case. Andreu and Vaughters both like Floyd Landis and would have felt uneasy about contributing in any way to the case against him. Vaughters and Landis are close friends. From all the text messages Landis received after his unexpected loss of ten minutes on stage sixteen of the 2006 Tour, it was Vaughters's that brought a smile to his face. Everyone else told him to see his collapse in perspective, it wasn't the end of the world; Vaughters said, "This sucks ass!"

Once parts of the IM conversation ended up in the SCA case, Vaughters provided an affidavit to Armstrong's lawyers saying that he exaggerated some of the things in the conversation and that some of the information had come from second- or thirdhand sources. However, on the key element dealing with the alleged blood doping by the Discovery team, he twice wrote in the instant message that his information had come from Landis. On the question of Landis, Andreu stresses that people should wait until both sides have fully presented their cases before deciding whether or not he deserves to be sanctioned. Vaughters says he does not believe Landis used testosterone to fuel his extraordinary ride on stage seventeen of the 2006 Tour but appears less sure of what Landis had done in the past. In a *New York Times* piece, published on July 30, 2006, he spoke to the journalist Ian Austen about how Landis was taking the news of his positive test. "More than anything," Vaughters said, "Floyd is very sad. I feel that he now may be thinking that this is retribution from God for all he did in the past."

Thirty-eight days before the start of the 2006 Tour de France, Spanish police raided three apartments and a medical clinic in Madrid.

They arrested five men from the community of professional cycling, and their investigation, code-named Operation Puerto, became the sport's latest drug scandal. Of the five taken into custody, Eufemiano Fuentes and Jose Luis Merino were doctors accused of running an elaborate sport doping network in Madrid. Manolo Saiz, directeur sportif with the Liberty Seguros team; Jose Ignacio Labarta, assistant directeur sportif with the Comunidad Valenciana team; and the mountain biker Alberto Leon were all accused of being clients of Fuentes and Merino. It was reported that the police found more than one hundred bags of frozen blood, and a stockpile of performance-enhancing drugs such as r-EPO, testosterone, anabolic steroids, and the controversial but legal substance Actovegin, which was found in U.S. Postal Service team waste at the 2000 Tour de France. The frozen blood was identified by code names that indicated connections with some of the sport's biggest names. According to documents seized by the police, fifty-eight professional cyclists were listed as clients of Fuentes and Merino.

Equipment taken from Fuentes's clinic showed the blood extracted from cyclists was cleaned and enhanced before being reinfused. Documents also proved that Fuentes was in charge of a multimillion-dollar business, with top riders paying anything from $60,000 to $100,000 each year for his services. Tyler Hamilton's first name figured prominently in documents leaked from Operation Puerto and the documents indicated that Hamilton was deeply involved with the Madrid clinic in 2003 and 2004. The most damning evidence against Hamilton is contained in a yearlong doping calendar, presumably devised by Fuentes or one of his associates, that purports to be the rider's drug regimen for 2003. There is also an equally suspicious document that reads as a fax sent to Haven Parchinski, which happens to be the maiden name of Hamilton's wife. It also correctly lists the fax number for the couple's apartment in Girona and, on another document, the fax number for their house in Boulder, Colorado. Hamilton denied any involvement with the Madrid clinic, said he had heard nothing from cycling officials about the affair but could not explain how such detailed information about him should have ended up in documents seized by the police.

Further documents leaked from the investigation indicated two of the sport's biggest names, the Italian Ivan Basso and the German Jan Ullrich, were among the fifty-eight riders doing business with Fuentes. This information was disclosed days before the start of the 2006 Tour de France, and, given the status of Basso and Ullrich, it created an obvious problem for the race organizers. The two riders were offered the opportunity to provide DNA samples that could be compared to the blood thought to be theirs in Madrid, but both preferred to plead their innocence while refusing to submit to DNA testing. They felt it violated their privacy. Pat McQuaid, Hein Verbruggen's successor as president of UCI, quoted the example of a murder investigation in London in which five thousand young men volunteered for DNA testing to eliminate themselves from the police inquiry. Including Basso and Ullrich, thirteen riders implicated in Operation Puerto were thrown out of the Tour de France.

Operation Puerto was much on everyone's minds as the 2006 Tour de France set off from Strasbourg on July 1. In one of his rare comments to the press, Dr. Fuentes said that if every rider who had come to him for help was barred from the Tour de France, the only ones who would make it to Paris would be the organizers. In all, three of the race favorites were kicked out—Basso, Ullrich, and the Spaniard Francisco Mancebo. That left Landis one of the favorites, but after losing ten minutes on stage sixteen to the mountaintop finish at La Toussuire, he produced what seemed a miraculous and heroic performance on stage seventeen to catapult himself back into contention. He then produced a strong time trial to win on the second to last stage to effectively win the race. Along the way, Landis was asked for his views on Operation Puerto and those caught up in the investigation. He was reluctant to offer an opinion. "Since you really want me to quote something about what happened at the beginning of the race, and you won't stop asking me questions, I'll say that it was an unfortunate situation for all of us, and none of us in any way got any satisfaction out of the fact that they're not here. Anybody got any questions about something else?"

By contrast, Bradley Wiggins, a young English cyclist who won a gold medal on the track at the Athens Olympics in 2004 and who was experiencing the Tour de France for the first time in 2006, was asked about those thrown out of the race. "It's good, bloody good. It's about time they got rid of those sods if they are proven to be doing stuff. Guys like me have to endure them making our lives hard in the mountains, and what happened is brilliant from that point of view."

Even before Landis's positive test, his Phonak sponsor had decided to leave the sport because of the numerous doping controversies that affected the team. Andy Rihs, the team owner, got a new sponsor, iShares, a subsidiary of Barclays Bank, but it withdrew after Landis's failed test. "Think hard before you get involved in cycling," said Rihs, "because there are never any guarantees when it comes to doping. Where there's money, there's doping." In the two years before Landis's positive, Phonak riders had been involved in a staggering number of doping controversies. Santiago Perez and Tyler Hamilton tested positive for homologous blood doping, Oscar Camenzind for r-EPO, Sascha Urweider for testosterone, Santos Gonzalez was dropped from the team following an internal doping control, and three of Phonak's riders—Santiago Botero, Jose Enrique Gutierrez, and Ignacio Gutierrez—were implicated in Operation Puerto. Then, there was Landis.

From this group, Oscar Camenzind stands out. The Swiss rider was a postman before realizing that his talent for bike racing could bring him greater rewards. He became one of the better riders of his generation and won the world road race championship in 1998. One April day in 2004, Camenzind was training near his home in Gersau, on the shores of Lake Lucerne, when he encountered drug testers from the Swiss Olympic Association. They asked if they could have a urine sample; he said they could but advised them that he had taken r-EPO and would test positive. Protocol demanded that the sample be given, and sure enough, it tested positive for r-EPO. Camenzind didn't bother with the B sample—rather, he admitted his guilt and retired from the sport. He returned to his life as a postman in Gersau.

Epilogue

THE MAN IS MORE THAN THE CYCLIST

The fifteenth leg of the 2000 Tour de France started in the town of Briançon and ended at the Alpine ski resort of Courchevel. Along the 108-mile route, the racers crossed two storied mountain passes, the Galibier and the Madeleine, before beginning the final ascent to the ski station at Courcheval, 2,004 meters above sea level. On days such as these, it is easy to use the word *epic* in describing the Tour de France. Against a staggeringly beautiful backdrop, eight breakaways push ahead of the pack, furiously making their way through the mountains, pursued by men content that time is on their side. After all, everyone knows the effort involved in forging a lead and then sustaining it. Almost imperceptibly, it drains a rider's energy. The breakaways would come back; they always did. But José Maria Jimenez was the desperado who believed he could avoid the inevitable. In his native Spain, Jimenez was renowned for his ability on mountainous roads and beloved for his aggressive style of racing. They called him Chava, short for *chaval,* Spanish for "young guy," and that was appropriate because Jimenez possessed the spirit of youth, and the soul of a man who didn't intend to age. And he soared like an eagle.

On the climb to Courchevel, the group of eight fragmented and not one of the other seven could hold Chava's wheel. Soon it was clear the threat came from behind, where a group led by Marco Pantani, Lance Armstrong, Richard Virenque, and Jan Ullrich was gaining fast. About three miles from the summit, Pantani accelerated in pursuit of Jimenez, whose lead was now just thirty seconds. Armstrong tried to go with Pantani but couldn't. Less than two miles from the finish, Pantani overtook Jimenez and that was how they rolled into Courchevel: first Pantani, then Jimenez, Armstrong fourth, and the rest of the pack in dribs and drabs. It had been one of those Alpine races that make fans on the mountainside feel lucky to have been witnesses. Two of the great climbers of this generation had given all that they had. And unknown to all, it was Pantani's final victory as a professional cyclist.

Fewer than three and a half years later, on December 6, 2003, José Maria Jimenez died in the bathroom adjoining his bedroom at the San Miguel psychiatric hospital in Madrid at the age of thirty-two. He had been suffering from depression for some time, and though it was said his death was caused by a heart attack, the truth was more painful. Chava died following a cocaine overdose. And, according to a nun who works at the clinic, the last hours were not easy for the racer who had won nine stages of the Tour of Spain and been the best mountain climber in the race from 1997 through 2001. One member of his family said that a cycling doctor prescribed cocaine to relieve the depression caused by steroid abuse. Earlier in his career, Chava had tested positive for caffeine and served a short suspension. For a time, he rolled with the punches and enjoyed the good life, but he was vulnerable—a vulnerable young athlete who happened to be in the wrong sport.

Less than a year before he died, José Maria Jimenez met Marco Pantani on one of the Canary Islands, and knowing each other from battles fought on murderous mountains, they shared a comfortable and mutual sense of empathy. They were different characters; Pantani saw himself as a champion—a much put-upon champion—while Jimenez had a more modest view of himself. Far from the gaze of their public, and the doping control officers who wanted their blood, it would be nice to think that they shared a fun evening together, because so much of their later lives were destroyed by cocaine addiction. Pantani's life ended nine weeks after Jimenez's.

Pantani's last week was grim. He booked into the modest Hotel Rose in the town of Rimini on Italy's Adriatic coast, spent much of the time in his room, appearing often enough for meals that when he didn't show on the Saturday, one member of the hotel staff became concerned. The police were called and they went to Pantani's room, which was locked from the inside. After gaining access, they found the 1998 Tour de France champion dead on the floor. Several packets of half-empty sedatives were found beside his bed, and a little over a month later, Dr. Giuseppe Fortuni delivered his report on the cause of death. "The death of Marco Pantani was caused by acute cocaine intoxication," he said. Marco Pantani and José Maria Jimenez had both used performance-enhancing drugs and they had both died from recreational drug use.

On the Saturday night of Pantani's death, the Italian anti-doping activist Professor Sandro Donati spoke of some of those who shared in the culpability for the tragedy. "This evening," he said, "there are doctors in Italy who should not sleep easily. Journalists, too, played a part. They knew what he [Pantani] was doing and they urged him to go faster and faster. When he won, they said he was a legend, when in fact he was very unhealthy." In the days after his death, Italian television played and replayed clips of Pantani's greatest rides in the mountains of the Tour de France and Giro d'Italia. Donati winced. "It was pornography, nothing else. There is meaning in Marco Pantani's life and death but we cannot find it in those pictures."

Pantani's funeral took place at the Catholic church in the small town of Cesenatico, where he had been baptized thirty-four years before. Yellow and white roses were placed on top of the coffin, alongside a photograph of the racer in the yellow jersey of the Tour de France. At the funeral service his manager and friend Manuela Ronchi read diarylike observations that Pantani had written into his passport in the weeks before his death. They reflected feelings of persecution and the bitterness he felt toward the cycling and police authorities who, he believed, had picked on him. "If I made mistakes, I'd like to know that there is proof, but when my sporting life, and above all my private life, was violated, I lost a lot. What is left? Just a lot of anger and sadness for the violence of the judicial system."

Speaking to the mourners, Bishop Antonio Lanfranchi asked that everyone look inside themselves. "Marco invites us to make a serious examination of our consciences, of everything that is sport and everything that is broken in sport. The man is greater than his victories and defeats. The man is worth more than the cyclist. In the champion beats the heart of a boy, a heart that needs normality and cannot be sacrificed at the altar of exploitation." How many others have perished at the altar of exploitation?

Five years after that thrilling race in 2000, the Tour de France returned to Courchevel. Barely eighteen months had passed since the deaths of the two supreme climbers, and yet if you had been on the mountain that afternoon, or read newspaper coverage of the race, you might not have guessed that they had existed. Cycling quickly forgets. Any attempt to find the reasons behind the deaths would have been too torturous—better not to go there. It has always been that way. In August 2005, the Tour de France organizer Jean-Marie Leblanc said the story he read about Lance Armstrong in *L'Equipe* proved to him that Armstrong used r-EPO to win the 1999 Tour and that he now believed the champion had fooled everyone. Eleven months later, on the day the first post-Armstrong Tour ended, Leblanc spoke about a recent conversation he'd had with the seven-time winner. "We explained ourselves well, and we all agreed that until proof to the contrary, Lance Armstrong has not lost the respect of the Tour de France. And he gives tons of respect to the Tour. It's sure we can't throw out Lance Armstrong's Tour de France years in a cloud of smoke." Had Leblanc forgotten the *L'Equipe* story and his unequivocal response to it?

Of course, it is easier to go with the flow and to see nothing but greatness in the legs of those who create daylight in the mountains, and it is better not to speak of the riders whose bags of enhanced blood still remain, unidentified, in the custody of Spanish police in Madrid. Is it simplistic to wonder why a suspected rider, such as the Italian Ivan Basso, would not travel to the Spanish capital and insist on being DNA tested?

Was it simplistic to wonder why the Discovery team would sign Basso when circumstantial evidence suggested he was one of the riders involved in Operation Puerto? In February 2007, the Discovery Channel announced that it would not be extending its deal with the team into 2008. Basso claimed to be unperturbed about that. "My deal is with Lance Armstrong, not Discovery," he said, a reference to the fact that it was Armstrong who had persuaded him to join the team. In April 2007, the Italian Olympic Committee (CONI) announced that it was reopening an investigation into Basso's involvement in Operation Puerto. A Spanish magazine, *Interviú*, reported that a diary containing suspicious information about Basso had turned up in the investigation of Eufemiano Fuentes, the central figure in the blood doping ring centered in Madrid. Among the tens of bags of blood found in Madrid, there are bags code-named "Birillo," which is believed to be Basso's dog's name. The impetus for the Italian investigation may have come from Germany, where prosecutors were able to match a DNA sample given by Jan Ullrich with blood seized in Operation Puerto. Once the match was confirmed, Ullrich retired from the sport.

For far too long, there has been the illusion of a serious fight against doping in cycling. But recent events—Operation Puerto in Spain and Floyd Landis's dramatic fall from grace—have focused minds like never before. Even Pat McQuaid, UCI president, speaks of a credibility problem for the sport, and his organization now promises to change rules to make DNA testing compulsory, and to target-test riders on four- or five-week training camps. It is a pity that these changes will come after many of the horses have bolted.

Perhaps there is hope in a new initiative, emphasizing the need for preventative testing rather than punitive testing. Jonathan Vaughters's United States–based Slipstream/Chipotle team has led the way with a system of internal testing that makes it extremely difficult for one of its riders to cheat and not get caught. In Europe, the T-Mobile team is clearly committed to playing its part with its own rigorous testing. One could also applaud the CSC team, as its Danish directeur sportif, Bjarne Riis, appears committed to serious change. His refusal to re-sign Basso, after the rider refused to submit to a DNA test, was courageous and correct. But Riis's support for a more aggressive stance against those suspected of involvement in doping raises an important question. Can cycling regain its credibility while it continues to hide its past?

Riis epitomizes the dilemma facing the sport. Everything that he now says encourages belief in his ethics, and yet it is hard to be certain. He won the 1996 Tour de France at a time when abuse of r-EPO was rampant, and his domination of that Tour was close to embarrassing. Can we ever forget that moment on the punishing climb to Hautacam when he peeled off the front of a select group of leaders, moved to the other side of the road, and slowed? As the others passed, Riis carefully assessed their states of tiredness before easing back into the slipstream of the last man. At first, it was thought he was in trouble, but that was soon followed by a sense of the opposite. He was so much in control that he could toy with his rivals. After taking those few seconds to evaluate his rivals' condition, he accelerated from last position, flew past them all, and just disappeared around the next corner. Next time they saw him was at the finish—the stage victory was his, and the Tour was as good as over.

The shadow of suspicion followed Riis for more than a decade. He was disparagingly known as Mr. Sixty Percent, and while he consistently denied ever having used r-EPO, the name stuck. In March

2006, a Belgian television investigation claimed his Telekom team in the 1990s used r-EPO and that the decision to dope was taken by the riders. Jef D'hondt, a soigneur with that Telekom team, accused German rider Uwe Ampler of introducing r-EPO to Telekom, and also alleged that Riis's particularly high hematocrit was caused by his use of the drug. "Riis had a hematocrit of sixty-four at one time during the Tour," said D'hondt (for a male athlete, forty-three to forty-five is the norm). Riis denied the new allegations. "I have never had a particularly close relationship with Jef D'hondt and he has no validation for the allegations he is making. To me, it's all in the past and I do not wish to be held accountable every time someone finds it interesting to bring up some ten-year-old story. I truly believe the future is much more important than the past. I want to be judged on the work I'm doing with my team today."

Patrick Lefevere, manager of the Quickstep Innergetic team, also appeared on the Belgian television program, and he wondered where it will all end. "Just what do you want, anyway? Do you want every rider to get down on his knees and confess?" Certainly not on their knees, and perhaps there is no need for confessions from every rider, but it is important to know how flawed the story has been. If Riis used r-EPO to help him win the 1996 Tour, it is right that we know it. The best future will be the one carved from the mistakes of the past. And how can there be trust in what Riis is now doing with the CSC team if he has not provided a complete account of his time as a rider? In the past, cycling's response to the Belgian TV investigation would have focused on Jef D'hondt and attempted to portray him as bitter, deceitful, vindictive, and utterly untrustworthy. No longer does that approach work, partly because so many have come forward, but also because it is now the sport that lacks credibility, not those who speak openly about doping.

As important as it is to know if Riis used r-EPO to win in '96, it is equally important to understand the recent history for American professional cycling. For eight straight summers, they played "The Star-Spangled Banner" on the Avenue des Champs-Elysées and it seemed the best of times. Now there is a fuller understanding of what was happening on the teams—U.S. Postal Service, Discovery, and Phonak—that delivered those eight successes. It is interesting to consider the Postal squad that helped Armstrong win his first Tour in 1999. Including Armstrong, there were seven Americans in the nine-rider team—Armstrong, Frankie Andreu, Tyler Hamilton, George Hincapie, Kevin Livingston, Jonathan Vaughters, and Christian Vandevelde. Very serious allegations have been made against Armstrong, especially *L'Equipe*'s claim that it has documents showing he used r-EPO for the 1999 Tour. Andreu has admitted he used r-EPO for the 1999 race; Hamilton tested positive for blood doping in 2004; documents seized by Italian police suggest Livingston used r-EPO in the late '90s; Emma O'Reilly has testified that she picked up testosterone for Hincapie from a contact at a Belgian hotel in 1998; and Vaughters has pointed the finger at himself. Writing in *The New York Times,* Juliet Macur quoted an unnamed rider from the 1999 Tour team. "The environment," he said, "was certainly one of, to be accepted, you had to use doping products. There was very high pressure to be one of the cool kids."

Vaughters is now an activist in the fight against doping and doing excellent work with the Slipstream/Chipotle team. But he has to only take a look at where Andreu now finds himself to know that things will not change easily. When Andreu admitted using r-EPO, he was criticized by his former directeur sportif Johan Bruyneel and also by UCI president Pat McQuaid. Though Andreu concentrated on his own experience and the key point was that the decision to use r-EPO was entirely his own, Bruyneel accused him of making "unfounded allegations." "I don't know what he's trying to achieve because he cannot achieve anything by saying this," said McQuaid. Perhaps Andreu felt that by explaining why he'd felt the need to dope, he was helping his sport. It may also be that he wished to answer a question honestly. Since retiring in 2000, Andreu has worked within the sport as a commentator, writer, and team manager. Since his admission, he has been less in demand. Yet the reaction from the public has been favorable, and Andreu has received more than a thousand e-mails and letters that have been overwhelmingly supportive. He still wants to work in the sport, and because of that, he doesn't know what the future holds.

When, in 2004, the former Motorola rider Stephen Swart admitted using r-EPO for the 1995 Tour de France, he was initially criticized in his home country. Jan Swart, Stephen's wife, read some of the things posted on message boards and heard other negative comments expressed on radio phone-ins, and they upset her. She asked her husband if he was sure he had done the right thing. "I reckon," he replied, "that if I live to be an old man, I will look back on my life and feel good about the fact that I told the truth about my cycling career." Swart wasn't surprised by people's reactions. "I think the people in New Zealand who know the sport, many of whom I know, didn't have their eyes opened [by the admission] because they were open already. There will always be a percentage of people who refuse to believe the reality. You're not going to change them."

The Swarts worried about the reaction of their teenage son, Logan. He rides his bike, likes the sport, and follows the Tour de France. Though he mightn't have said so, it was clear Logan was proud of the fact that his dad had ridden and finished the Tour. Stephen did not tell his son about the sport's doping culture and did not speak to him about his own use of performance-enhancing drugs. Before it came out in *L.A. Confidentiel,* Logan knew nothing. But Stephen and Jan need not have been concerned. When they spoke with Logan, it was soon apparent that he had nothing but respect for his dad's honesty and appreciated knowing what it had been truly like for his father. "I think Logan now sees the real picture," Stephen said.

Swart was last heard from in March 2007, at the end of the first day of a weeklong off-road bike tour of New Zealand's South Island. Twelve riders took a flight from Auckland on the North Island to Dunedin, boarded a train to Middlemarch, then biked it to Ranfurly. From there, they would ride to Alexandra and then another bike ride to Queenstown. If you've been to New Zealand, or even if you have seen any of the *Lord of the Rings* films, you will have a sense of the week they had. He talked about the scenery on that first ride to Ranfurly, and said he had never seen anything like it in his life. He sounded like a boy enjoying his first summer camp.

Emma O'Reilly got out of professional cycling at the ripe old age of thirty, as she had planned to do. Five years on the road with the U.S. Postal Service team was as much as she wanted, and though it pleased her to have risen to head soigneur on the team, she left without looking back. There was a real world out there—a place to which she was glad to return. She agreed to do an interview with the authors of the book *L.A. Confidentiel* because she didn't see why not. It wasn't like she had been involved in the medical program. Seeing from inside the team, she had an insight and an experience that she thought might highlight the need for change within the sport—although her personal view is that not much will change. When the book was published and her name appeared in newspaper headlines, TV news summaries, and numerous cycling publications and on websites, she cringed. At no time had she anticipated that kind of attention, and when it came, it upset her.

Another irritation was the subpoena that forced her to testify in the SCA case. At different times in that case, Armstrong's attorney Timothy Herman suggested, both to her and to other witnesses, that she had been involved in relationships with some of the riders and that she had a problem with alcohol. She never realized a lawyer could base questions on fabrications, but afterward, she was told it happened all the time when one side was attempting to destroy the credibility of the other side's witness. Little did she realize what she was getting into when agreeing to be interviewed by the authors of *L.A. Confidentiel.* They told her she would be one of many witnesses quoted in the book, but it was her interview that lasted seven hours. She thought that would be the end of it—it was only the beginning. They sent her the transcript of the interview, which ran to seventy-four tightly spaced pages, and asked if she would take the trouble to read it carefully and make whatever changes she thought appropriate. So every evening for two weeks after work at the medical center in Manchester, she worked on that transcript until she got it exactly as she wanted it. Then they wanted her to read the chapters that covered her part in the story, and those chapters were about a quarter of the book.

She had been fine about giving the interview but this was too much. Why should she have to spend so much time helping authors who were being paid for their work? It wasn't like they were donating all proceeds to charity. After she read the first drafts of those chapters, they asked if she would mind reading the final drafts. And afterward, would she mind doing just two or three interviews to help publicize the book? So she told them she felt she was being exploited and suggested she should be paid for the time she was spending on the book. They agreed to give her 2.5 percent of what they each earned from the book, which came to £5,000 for her. The money wasn't a lot but at least it was recognition for the work she had done. Armstrong's lawyer Timothy Herman asked her about it during the SCA case and she told him it was the hardest £5,000 she had earned in her life.

O'Reilly now says if she could go back in time, she wouldn't do the interview. After the deaths of Marco Pantani and so many other young cyclists, she wanted to help change things by telling the truth about the U.S. Postal Service team. The reaction to the book convinced her that the sport's leaders didn't really want things to change. She can't see what good came from her honesty, but at least she now understands why so many observe the sport's law of silence. Eventually, her life returned to normal and she concentrated on her career at the Physical Therapy Clinic in Hale, Manchester, and when an opportunity came to buy the business, she did. It's now called the Body Clinic and the practice has become bigger and more popular under her ownership. Her life in cycling was good. What has happened since is better.

Many who followed the Lance Armstrong story believed the truth would emerge in a no-holds-barred court case. They, too, were to be disappointed. Armstrong's lawsuit against SCA Promotions went on for more than a year and resulted in an out-of-court settlement in his favor—he received the $5 million bonus due to him and a further $2.5 million in costs. His victory was assured once Richard Faulkner, chairman of the arbitration panel, ruled that SCA was an insurance company. Once that was decided, all that mattered was that the UCI considered Armstrong the winner of the 2004 Tour. His libel actions weren't so conclusive. After trial dates were agreed in both London and Paris, Armstrong settled in one city and withdrew in the other. The settlement with *The Sunday Times* in London was a victory of sorts for Armstrong because the newspaper agreed to apologize for the offending article and to pay one third of his legal costs. Under France's more sympathetic libel laws, the publishers of *L.A. Confidentiel* let it be known they were not for settling. Shortly before the trial was scheduled to happen in October 2005, Armstrong withdrew his action. Bill Stapleton's prophecy, uttered during the secretly taped conversation with Frankie Andreu at the 2004 Tour de France, had come to pass. "Because the best result for us is . . . drop the fucking lawsuit and it all just goes away. Because the other option is full-out war in a French court and everybody's gonna testify and it could blow the whole sport."

AUTHOR'S NOTE

"Hope is nature's veil for hiding truth's nakedness."
—Alfred Nobel

Four years ago, I traveled to Milan to meet a young American who had recently moved from Colorado to Italy. He told a story about a friend of his, a European-born professional cyclist, who had asked this young man to bring to Italy a pair of favorite cycling shoes he had unintentionally left in the United States. The shoes were dropped off at the young man's house in Boulder—a friend of the owner of the shoes just left the package at this young man's door. As it turned out, the traveler didn't have much spare room, but he figured he could squeeze in the shoes if he took them out of the package and packed the two shoes separately.

Inside the package were eight cartons of bovine hemoglobin. (It was the spring of 2003—bovine hemoglobin was still "in.") The Coloradoan wasn't certain what to do, but playing the mule wasn't high on his list of career ambitions. He called a neighbor, a university professor, and they decided to pour the hemoglobin down the sink. After arriving in Italy, he had it out with his so-called friend, who apologized for deceiving him about the bovine hemoglobin. But the cyclist's contrition disappeared as soon as he heard that the contents of the package had been drained. "Do you realize how much that stuff cost?" he shouted.

Though the young man told me the cyclist's name, he did not want the guy publicly shamed, mostly because of the attention that would fall on the whistle-blower. I follow the rider's performances each year, and, over the last three seasons, he has become quite a star. He has won stages of the Tour de France and is expected to claim another in 2007. Journalists write about him as a man and athlete we should admire.

I don't blame my informant for not wanting to publicly name the rider. Those who tread on the toes of star athletes and on the dreams of their fans receive no thanks and little mercy. It is why I have so much admiration for former racers such as Steve Swart and Frankie Andreu, who have sacrificed the friendship of some former teammates, and others within professional cycling, to tell the truth about their own careers and the environment that virtually coerced them into using r-EPO.

Swart and Andreu were just two of many sources who helped in putting together this story. Like others, they put themselves out on a limb. Greg LeMond and Emma O'Reilly had no inkling of the opprobrium to which they would be subjected when they agreed to speak honestly, and I am sure they have since wondered if it was worth the hassle. Others were just as brave. I am thinking especially of Dr. Prentice Steffen, Dr. Michael Ashenden, and Betsy Andreu. Their only aim was that the truth would come out.

Many others agreed to be interviewed: the former Motorola riders Phil Anderson and Andy Hampsten; the former U.S. Postal Service riders Jonathan Vaughters and Marty Jemison; the former Motorola directeur sportif Mike Neel; and the former Festina directeur sportif Bruno Roussel. Others were equally generous in many different ways—Piero Boccarossa, Dean Brewer, Eugenio Capodacqua, Sandro Donati, Les Earnest, Thom Gunn, Vivian Hackman, Andrew James, Jorge Jasson, Lars Jorgensen, Joe Lindsey, Josie, Gwen Knapp, Kathy LeMond, Stephane Mandard, Stephanie McIlvain, Dawn Polay, Charles Pelkey, John Pineau, Dr. Greg Strock, Phil Taylor, Dr. Max Testa, Lory Testasecca, Antoine Vayer, René Wenzel, and Peter and Kristine Zaballos. Most interviews were done face-to-face, using a tape recorder, and many of the sources would later retell their stories under oath in the case taken by Lance Armstrong and Tailwind Sports against the Dallas-based company, SCA Promotions.

There were many reasons for writing this book, but the greatest was the desire to support the numerous cyclists who wish to race clean. More than anything, they are entitled to know how good they are and to have a shot at realizing their true potential. One understands why they remain a relatively silent majority, but it is a strange world in which clean racers feel that they cannot be perceived to be vigorously against doping. There have been exceptions. My great friend, Paul Kimmage, wrote the story of his life as a professional cyclist, *Rough Ride,* and it remains the definitive account of a sport gone badly wrong. With the skillful help of Benoît Hopquin, Christophe Bassons wrote a similarly absorbing book ten years later, *Positif* (Positive), and, perhaps not surprisingly, Bassons's experience was that the doping culture was as deep-rooted in his time as it was in Kimmage's. Pierre Ballester is another who spent a lot of time on the antidoping trail, and it was a pleasure to spend some of the journey in his company. We have forever lived with the hope that professional cycling will somehow cure itself.

It is customary for an author to thank his agent and publisher, and I sense that sometimes it is perfunctory gratitude. Not so in this case. Without Scott Waxman's intelligence and Random House's willingness to take on the challenge, this story might never have been told. For this European, Random House offered proof that America remains a great democracy. Not only that, but the publisher provided Mark Tavani, who was a skillful and patient editor. From the beginning, he believed in the authenticity of the story. Thanks are also due to Mark's assistant, Paul Taunton. Bill Adams, too, did a great job. How many people are human beings first and lawyers second? Bill was one of the exceptions. Thanks also to Elizabeth McGuire, publisher of Ballantine; Cindy Murray, publicist; Brian McLendon and Carol Schneider, directors of publicity; Claire Tisne, Rachel Kind, and Rachel Bernstein in Subrights; and Kim Hovey, director of marketing. I would also like to thank some great people from *The Sunday Times* in London who have been staunchly supportive through the years. I am thinking, especially, of Alex Butler, Richard Caseby, and Gill Phillips. And how could I forget the unflinching support of Alan English?

To my own family, who have lived with this project for as long as I have, I say a heartfelt thank you.

ABOUT THE AUTHOR

DAVID WALSH is chief sportswriter with *The Sunday Times* (London). A four-time Irish Sportswriter of the Year and a three-time U.K. Sportswriter of the Year, he is married with seven children and lives in Cambridge, England. He is co-author of *L.A. Confidential: The Secrets of Lance Armstrong.*

ABOUT THE TYPE

This book was set in Bodoni Book, a typeface named after Giambattista Bodoni, an Italian printer and type designer of the late eighteenth and early nineteenth century. It is not actually one of Bodoni's fonts but a modern version based on his style and manner and is distinguished by a marked contrast between the thick and thin elements of the letters.

BALDWIN PUBLIC LIBRARY
3 1115 00559 1986